FORMAL EQUIVALENCE CHECKING AND DESIGN DEBUGGING

FRONTIERS IN ELECTRONIC TESTING

Consulting Editor
Vishwani D. Agrawal

Books in the series:

On-Line Testing for VLSI
M. Nicolaidis, Y. Zorian
ISBN: 0-7923-8132-7
Defect Oriented Testing for CMOS Analog and Digital Circuits
M. Sachdev
ISBN: 0-7923-8083-5
Reasoning in Boolean Networks: Logic Synthesis and Verification Using Testing Techniques
W. Kunz, D. Stoffel
ISBN: 0-7923-9921-8
Introduction to I_{DDQ} Testing
S. Chakravarty, P.J. Thadikaran
ISBN: 0-7923-9945-5
Multi-Chip Module Test Strategies
Y. Zorian
ISBN: 0-7923-9920-X
Testing and Testable Design of High-Density Random-Access Memories
P. Mazumder, K. Chakraborty
ISBN: 0-7923-9782-7
From Contamination to Defects, Faults and Yield Loss
J.B. Khare, W. Maly
ISBN: 0-7923-9714-2
Efficient Branch and Bound Search with Applications to Computer-Aided Design
X.Chen, M.L. Bushnell
ISBN: 0-7923-9673-1
Testability Concepts for Digital ICs: The Macro Test Approach
F.P.M. Beenker, R.G. Bennetts, A.P. Thijssen
ISBN: 0-7923-9658-8
Economics of Electronic Design, Manufacture and Test
M. Abadir, A.P. Ambler
ISBN: 0-7923-9471-2
I_{DDQ} Testing of VLSI Circuits
R. Gulati, C. Hawkins
ISBN: 0-7923-9315-5

FORMAL EQUIVALENCE CHECKING AND DESIGN DEBUGGING

by

Shi-Yu Huang
National Semiconductor Corporation

and

Kwang-Ting (Tim) Cheng
University of California, Santa Barbara

KLUWER ACADEMIC PUBLISHERS
Boston / Dordrecht / London

Distributors for North, Central and South America:
Kluwer Academic Publishers
101 Philip Drive
Assinippi Park
Norwell, Massachusetts 02061 USA

Distributors for all other countries:
Kluwer Academic Publishers
Distribution Centre
Post Office Box 322
3300 AH Dordrecht, THE NETHERLANDS

Library of Congress Cataloging-in-Publication Data

A C.I.P. Catalogue record for this book is available
from the Library of Congress.

Printed on acid-free paper.

Printed in the United States of America

FORMAL EQUIVALENCE CHECKING AND DESIGN DEBUGGING

CONTENTS

Foreword

The world is increasingly dependent on electronic systems. Electronic systems have requirements on speed, cost, power, reliability, and on the correctness of their functioning. Of these requirements functional correctness is the most fundamental requirement because the speed and reliability of an incorrectly functioning electronic system is of no interest. To function correctly circuits must be designed, implemented and manufactured correctly. Verifying functional correctness is also becoming the single largest bottleneck in the design process and to understand where that time and effort is going it will be useful to further explain the nature of these verification problems.

Given a specification of the functionality of an integrated circuit (IC), *design verification* is the verification that the design of an IC is consistent with its specification. By the design of an IC we mean the description which embodies the designer's intent. For example in a register-transfer level (RTL) synthesis methodology, the design is initially captured as an RTL model of the circuit in a hardware-description language (HDL). This RTL model, typically coded in Verilog or VHDL, can be both simulated and synthesized. The design verification process typically proceeds by creating a simulation driver which produces simulation stimuli for the RTL model and a simulation monitor which monitors the correctness of the outputs. This basic structure of the simulation model, driver and monitor has remained unchanged for well over a decade. Progress in simulation has principally focused on either raising the level of abstraction of simulation (e.g., from gate-level to RTL) or improving the speed of simulation (e.g., from interpreted to compiled or cycle-based). Much research effort has gone into exploring ways to apply formal verification techniques, such as model-checking, to the design verification problem, but to date the success of formal design verification has been limited to niche applications. Perhaps the most successful use of formal design verification techniques will be in using formal models to augment and direct simulation. Be that as it may, simulation technology continues to remain the predominant approach for verifying the correctness of the initial design entry.

At the other extreme of the verification process is manufacture verification. After the production of the physical layout and masks from the implementation process, *manufacture verification* is the verification that the manufacturing process has not introduced defects during the production of the actual IC. This is more commonly referred to as *manufacture test* or simply *test*. The manufacture test process is similar to the design verification process. A driver (the tester) is created which delivers the stimuli (test vectors) for the IC and a monitor (also the tester) is created which monitors the correctness of the outputs. The major difference is that the stimuli are applied to the actual IC as opposed to a model of the IC and as a result the test equipment must be capable of running at very high speeds.

Implementation stands between the entry of a design of an IC and its manufacture. Implementation constitutes the stages of refinement through which an initial description of a design, such as an RTL-model, becomes gradually transformed and elaborated into a physical layout (see Figure 1.2). Implementation verification is the process of verifying that the physical layout of the IC is consistent with the initial design entry. For example, in an RTL-synthesis methodology this verifies that the steps of logic synthesis, logic optimization, test synthesis and physical design have produced a physical layout that is consistent with the initial RTL model of the circuit. In all verification steps all design characteristics need to be verified but this is especially true in implementation verification. The most commonly verified design characteristics are functionality and speed. The most common approach to implementation has been simulation.

For example, in an RTL-synthesis methodology, simulation is used to verify that the gate-level netlist is equivalent to the initial RTL model. As before, simulation drivers and monitors are constructed to verify proper functionality. As the resulting gate-level model is modified, either due to manual intervention or due to tools such as scan-test-insertion, gate-level simulation is used to verify that the proper functionality and performance are maintained. Finally, after place and route tools translate the gate-level netlist into an actual circuit layout, two additional implementation verification steps are performed. The first is the verification of the circuit layout against the gate-level schematic. This is a structural check. Then wiring capacitances are extracted and back-annotated onto the gate-level netlist. With this more accurate information regarding wiring-delay, the gate-level simulation is run again to verify that the circuit speed is still within allowable margins.

Traditionally, gate-level simulation has been so central to the verification of the function and delay of a circuit that the gate-level simulator has been called the "golden simulator." It is called "golden" because it establishes the "gold standard" for the speed and functionality of the circuit. Indeed, there are many advantages to using gate-level simulation for these diverse implementation verification problems. The first is that the use of simulation is well understood by designers. Also, simulation allows the designer to simultaneously verify both the speed and functionality of a circuit. Using simulation as the workhorse for implementation verification has numerous disadvantages as well. The first of these is the growing computation time required to complete the verification of the implementation of a sizable (> 1 million gates) IC. The second is the lack of completeness. The verification of the correctness of the implementation of an IC is limited to those vectors that were applied. If the vector set was inadequate the functionality of the IC may be incorrect or the speed of the IC may be inadequate.

Fortunately, two complementary technologies have matured to replace gate-level simulation for the problem of implementation verification: *static timing verification* and *equivalence checking*. The resulting methodological approach is called *static sign-off* because strictly static techniques are used to verify, or "sign-off" on an IC. Static timing verification is a technology that implicitly analyzes the delay of all paths of a circuit and determines the worst-case delay without requiring any input stimuli. This allows an alternative approach to gate-level simulation for verifying the speed of a circuit in the successive stages of implementation. As a technology, static-timing verification is over twenty years old and this technology has gradually gained commercial acceptance. This still leaves the problem of verifying the functionality of a circuit, which is addressed by equivalence checking.

Equivalence checking attacks a fundamentally hard problem: Do two sequential circuits show equivalent functionality over all input stimuli? Progress on formal techniques for verifying equivalence has been steady and we are just now reaching the point where commercially available approaches to this problem are establishing a beachhead among designers. For this reason the publication of this book is very timely. The implementation verification problems that equivalence checking addresses are well motivated and illustrated in Chapter 1 and Part I as a whole does an admirable job of reviewing the design problems that require equivalence checking and describing the underlying technologies that are used to solve them.

One of the remaining technological challenges in the use of equivalence checking is the automated verification of sequential circuits that have equivalent external behavior but have different internal register placement. Finite-state assignment tools may modify the internal state representation of a sequential circuit while retaining the circuit's input/output behavior. Similarly, retiming optimizations, in which registers are moved forward or backward in the circuit to improve timing pose similar challenges to equivalence checking tools. The final chapter of Part I, Chapter 6, gives some novel approaches to these problems.

Imagine the frustration of a designer who has been methodically working through the implementation of a circuit for months. Modification after modification has been integrated into the design and verified. Then after a particularly intensive series of design modifications, equivalence checking is run again. Equivalence between this design iteration and the last is expected, but instead the circuits are found to be different. Where did the error occur? Was it the recent performance tuning? The scan-chain reordering? Perhaps an inverter was inadvertently used instead of a buffer? In actual tool usage, designers are apt to spend more time and effort diagnosing errors than verifying equivalence. However, academic research has focused on how to efficiently verify the equivalence of two circuits and has largely neglected the question of how to aid the designer in tracking down errors when two circuits are found not to be equivalent. A strength of this book is the breadth of coverage of the problem of error diagnosis in Part II.

With the adoption of the *static sign-off* approach to verifying circuit implementations the application-specific integrated circuit (ASIC) industry will experience the first radical methodological revolution since the adoption of logic synthesis. Equivalence checking is one of the two critical elements of this methodological revolution. This book is timely for either the designer seeking to better understand the mechanics of equivalence checking or for the CAD researcher who wishes to investigate well-motivated research problems such as equivalence checking of retimed designs or error diagnosis in sequential circuits.

Kurt Keutzer
University of California, Berkeley

Preface

Verification and debugging are important design activities for today's complex electronic designs. The enormous cost and time required to validate the designs has led to the realization that more powerful and efficient methodologies and techniques must be developed and adopted. In both academia and industry, efforts and interest in the development of formal methods in this area has increased dramatically during the past few years. These efforts have resulted in a number of exciting approaches and several new commercial tools.

This book covers two major topics in design verification: logic equivalence checking and design debugging. The first part of the book reviews the role of logic equivalence checking in the design process and describes the underlying technologies that are used to solve this problem. Chapters 2 and 3 provide a concise introduction to two fundamental techniques for logic equivalence checking: symbolic verification and incremental verification. Chapters 4 and 5 report some of the results of our recent research in this area. A software prototype system named AQUILA has been developed based on the ideas presented in these two chapters. Some novel approaches to the problems of verifying retimed circuits and verifying a gate-level implementation against its RTL specification are introduced in Chapters 6 and 7. The second part of the book (Chapters 8 though 11) provides an introduction to the problems of design error diagnosis and design error correction, a thorough survey of previous and recent literature, and an in-depth explanation of some of our recent research results on this topic. The algorithms described in Chapters 9, 10 and 11 have been implemented as two prototype programs, ErrorTracer and AutoFix, for design debugging.

The authors' interest in symbolic methods owes much to the discussion with Dr. Kuang-Chieh Chen of Verplex Systems Incorporated. and Professor Malgorzata Marek-Sadowska of University of California, Santa Barbara. Special gratitude is due to Dr. Vishwani Agrawal of Bell Labs., Lucent Technologies for his thorough review of the entire manuscript. Also, we want to thank Professor Forrest Brewer, Dr. Uwe Glaeser, Dr.

David Ihsin Cheng, and Dr. Chih-Chieh Lin, for their technical contributions to this research. Most of the research described in this book was conducted under the support from National Science Foundation grant MIP-9503651, California MICRO, Fujitsu Labs of America, Rockwell International Corporation, and National Semiconductor Corporation.

Shi-Yu Huang
National Semiconductor Corp.
Santa Clara, California

Kwang-Ting (Tim) Cheng
University of California
Santa Barbara, California

Chapter 1

Introduction

1.1 Problems of Interest

Hardware verification is a process of checking if a design conforms to its specifications of functionality, timing, testability, and power dissipation. Among these criteria, functional verification has the highest priority in order to guarantee the functional correctness of the design. As the complexity of a VLSI design grows, verification quickly becomes the bottleneck in the design process. It has been reported that more than half of the total design effort was devoted to the verification and debugging process for a high-performance microprocessor design [67]. Functional verification methods can be classified into three categories: software simulation, hardware emulation, and formal verification as shown in Fig. 1.1.

Simulation has been the most common way for functional verification. However, that approach quickly runs out of steam as the design becomes larger. Simulation has two major limitations. First, it is quite ad-hoc and not complete, and thus, may fail to catch some design errors. Secondly, it is very time-consuming when performed at the lower levels of abstraction, e.g., the gate-level or the transistor-level. Simulation is an adequate method

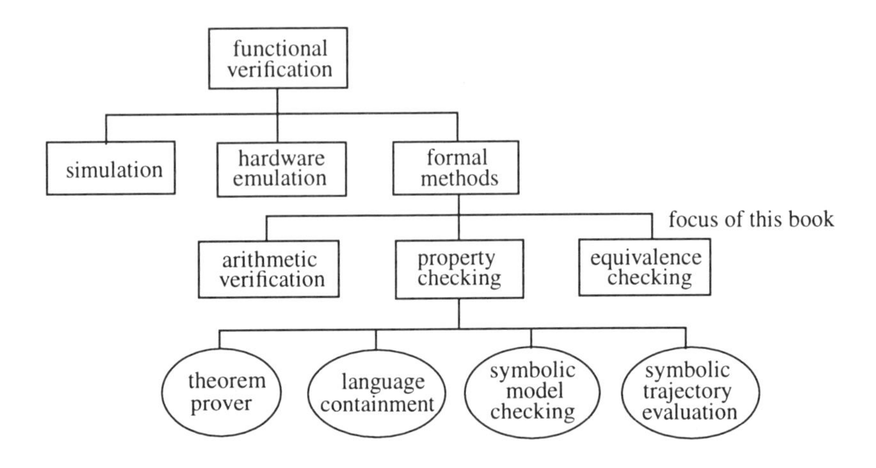

Fig. 1.1 A dichotomy of functional verification.

for developing manufacturing tests [30]. However, it is used for design verification only in the absence of a better method.

The technology of hardware emulation using Field Programmable Gate Array (FPGA) chips is gaining some popularity and provides a way to speed up the simulation process by several orders of magnitude. The downside of this approach is that hardware emulators remain expensive and it may take a long time to map the design under verification to the emulator.

On the other hand, formal verification has emerged as another solution to prove the correctness of a design. For the last decade, a wide spectrum of formal verification methods has been proposed. Most of them can be classified into three categories:

- *Arithmetic verification*: this proves an implementation is indeed performing the desired arithmetic operation, e.g., a multiplier is indeed performing the multiplication [19].
- *Property checking*: this proves that the design satisfies a property, e.g., proving a design realizing a communication protocol is deadlock-free. Property checking can be further divided into several paradigms, e.g., theorem proving [67], model checking [21,41], language

containment [81], and symbolic trajectory evaluation [20]. Each of these paradigms provides a unique formalism for specifying the property to be checked, and a formal procedure to verify the property. Applications of formal methods to validation of computer protocol specifications and to conformance testing of protocol implementations have been described by Holzmann [57].

- *Equivalence checking*: this proves that two given designs have the same functionality, e.g., an optimized design is functionally equivalent to its earlier version. Equivalence checking is sometimes referred to as *logic verification, implementation verification*, or *Boolean comparison*. We use these terms interchangeably in the sequel.

1.1.1 Equivalence Checking

In the first part of this book we focus on the problem of equivalence checking. Equivalence checking can be applied at different stages during the design process. Fig. 1.2 shows the role of equivalence checking in a typical flow of modern VLSI design. In this flow, the design is either specified at the register-transfer level using a hardware description language such as Verilog or VHDL, or described at the micro-architecture level. This high-level specification is then synthesized to a gate-level netlist. Next, a sequence of optimization steps is performed to improve the area, timing, testability, or power dissipation. The optimized implementation is then mapped using a given cell library through technology mapping. After that, the physical design follows.

During the design process, checking the equivalence of two designs described at the same or different levels of abstraction is necessary. For example, checking the functional equivalence of the optimized implementation against the RTL specification is critically important in order to guarantee that no error is introduced during the logic synthesis and optimization process, especially when human effort is involved in the process. Similarly, checking the equivalence of the gate-level implementation versus the gate-level model extracted from the layout can assure that no error is made during the physical design process.

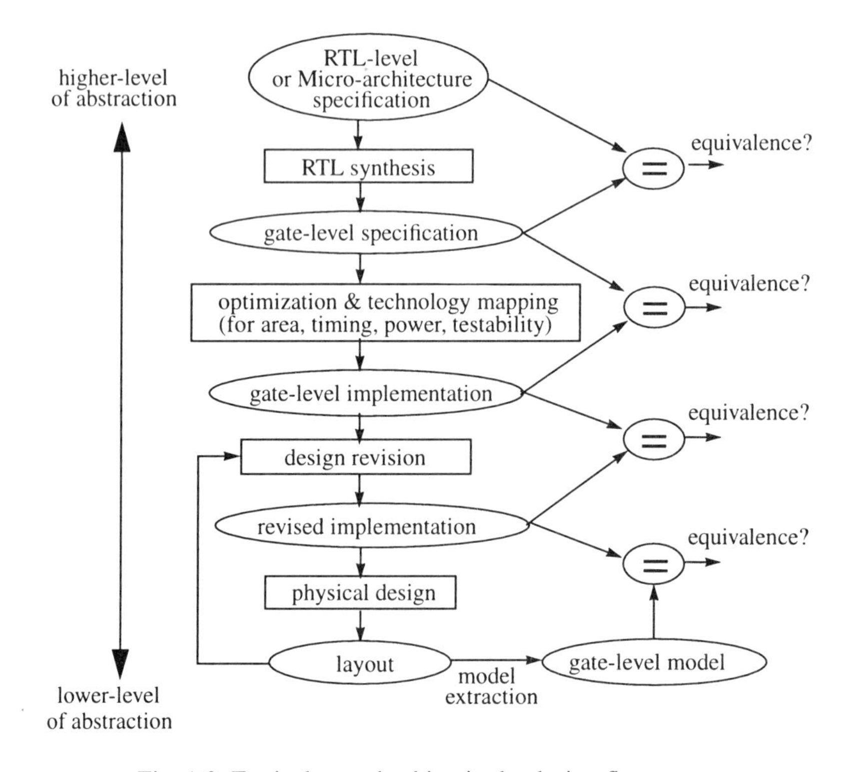

Fig. 1.2 Equivalence checking in the design flow.

Fig. 1.3 shows a taxonomy of the formal methods for equivalence checking. In this classification, the verification approaches are divided into two general categories: (1) symbolic approaches, and (2) incremental approaches. Symbolic approaches are referred to as those approaches that rely heavily on symbolic techniques using the Ordered Binary Decision Diagram (OBDD) [18,37,42,55,59,124]. An alternative form of symbolic approach relies on logic programming languages like Prolog [122]. Incremental approaches are those approaches that explore the structural similarity of the two circuits under verification (CUVs) [13,62,63,71,78,97,110,112]. The incremental approaches can be further classified into three types: (a) substitution-based [13,62,75,97],

(b) learning-based [71,78,112], and (c) transformation-based [110] algorithms. In comparison, the incremental approaches, although they rely

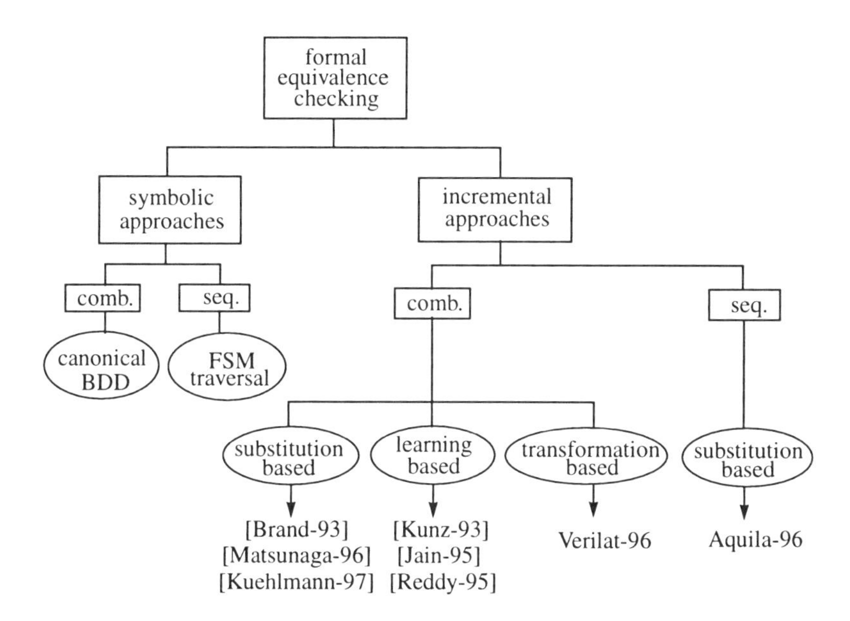

Fig. 1.3 A taxonomy of formal methods for equivalence checking.

on the structure similarities between CUVs, have empirically demonstrated their capabilities in handling larger circuits after intensive logic optimization. On the other hand, the BDD-based symbolic approaches are independent of the CUVs' structures and, thus, are more suitable for verifying two control-dominant circuits that are differently encoded and/or have completely different structures.

Traditionally, checking the functional equivalence of two combinational circuits is accomplished by constructing their canonical representations, e.g., truth-tables, or BDDs [8,18]. Two circuits are equivalent if and only if their canonical representations are isomorphic. To verify the equivalence of two sequential circuits, the circuits under verification (CUVs) are often regarded as finite state machines (FSMs). Assuming that two machines have a known reset state, they are equivalent if and only if their reset states are equivalent.

A product machine of two CUVs can be formed by connecting each corresponding primary input pair of the two machines together, and connecting each corresponding primary output pair to an XOR gate as shown in Fig. 1.4. This model is also referred to as the *miter* [13,44]. The

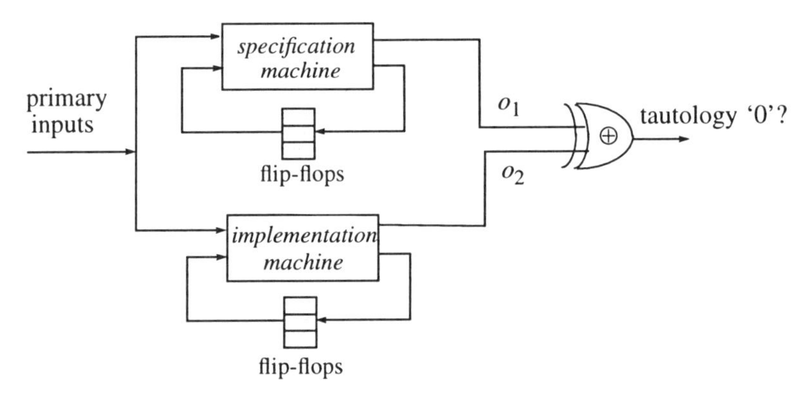

Fig. 1.4 Product machine (miter).

outputs of these XOR-gates are the new primary outputs of the product machine. The equivalence of the two machines can be asserted if every primary output of the product machine is tautology '0' for any input sequence. That is, the primary output response of the product machine is always '0' for any input vector and reachable state. Usually, the first step in proving machine equivalence is to compute the reachable states of the product machine from the reset state. This procedure is based upon finite state machine traversal.

Although the finite state machine traversal techniques have progressed over the last decade significantly through the advances in the BDD techniques, they still fail on many practical designs because the construction of the required BDD representations may cause memory explosion.

To overcome the limitations of the symbolic approaches, an algorithm that performs the combinational verification in stages was proposed by Brand [13] in 1993. This algorithm explores the structural similarity of the two CUV's to speed up the verification process and has successfully verified larger circuits. A similar algorithm was also proposed by Kunz and Stoffel [79] around the same time. These works sparked great interest in incremental algorithms for equivalence checking. Several other techniques were further proposed to either improve the efficiency or enlarge the applicable domain [62,71,97,110,112]. An alternative approach to equivalence checking defines a *characteristic polynomial* for the Boolean function. Although the polynomial is not a unique representation of the Boolean function, its evaluation allows an equivalence check in a probabilistic sense [5,6,70].

However, these techniques are only for combinational circuits. When comparing an RTL specification with a hand-crafted netlist, there may not exist a one-to-one flip-flop correspondence between the two designs, and thus, these approaches cannot be applied. Also, state-of-the-art logic optimization tools may apply sequential transformations such as retiming [90], sequential redundancy removal [32], and redundancy addition and removal [33]. These transformations may alter the functionality of the combinational portion of a circuit while preserving the overall sequential input/output behavior. It has been demonstrated that these transformations can handle circuits with hundreds or even thousands of flip-flops and may remove flip-flops from the original design. Even though these approaches are theoretically correct, the tools that implement the algorithms may not be error-free. Making sure that the transformed circuit indeed preserves the input/output behavior of the original circuit requires sequential verification.

In the first part of this book, we first review the fundamental approaches to equivalence checking (Chapters 2 and 3). Then we introduce a framework to generalize the incremental verification algorithms to sequential circuits (Chapters 4, 5, 6, and 7). The key features of this framework include: (1) an algorithm to verify circuits with or without a known reset state. (2) an inductive algorithm for exploring sequential similarity between the two CUVs, (3) a heuristic for verifying the circuit after a retiming transformation, (4) a formulation that can take external don't care information into account, and finally (5) an algorithm that integrates the FSM traversal techniques with the incremental verification approaches for verifying complex designs with a significant amount of data-path circuitry as well as differently encoded controllers. Experimental results have demonstrated that this framework is suitable for many circuits that are beyond the reach of other existing approaches.

1.1.2 Design Error Diagnosis and Correction

During the VLSI design process, functional mismatches between a given specification and the final implementation often occur. Once a functional mismatch is found by the verification tool, the designer faces the daunting task of design error diagnosis – a process that identifies or narrows down the error sources in the implementation, so as to assist the subsequent error correction process.

Most existing approaches for design error diagnosis fall into two main categories: the BDD-based methods and the simulation-based methods. The BDD-based methods [40,87,88,92,126] try to find those signals that are most likely to be responsible for the output incorrectness. A set of signals is *totally responsible* for the incorrectness if there exists a new function for each signal so that the implementation can be completely fixed. In other words, one can correct the entire implementation by re-synthesizing these signals. On the other hand, the simulation-based methods [73,108,109,123] are based on some heuristics to narrow down the locations of the errors. Usually, a number of input vectors that produce erroneous responses are simulated to narrow down the potential error region gradually. Fig. 1.5 compares these two methods in terms of the accuracy and the size of the applicable circuits. It shows that, the BDD-based approaches are highly accurate in identifying the error sources. On the other hand, the simulation-based approaches, although less accurate, scale well with the circuit size. In Chapter 8, we will discuss these algorithms in detail.

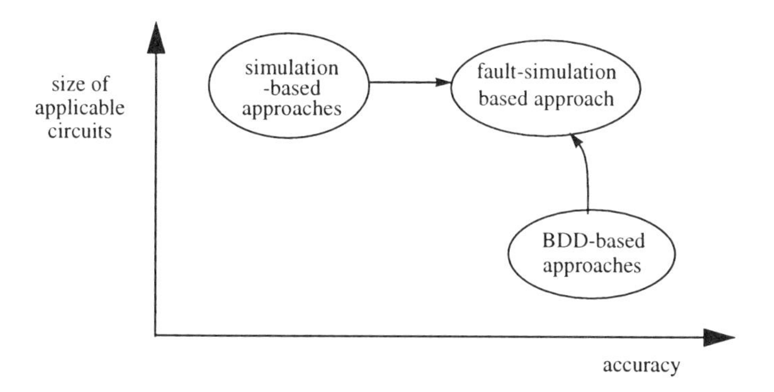

Fig. 1.5 Comparison of design error diagnosis approaches.

In Chapters 9 and 10, we introduce a new formulation for design error diagnosis based on fault simulation techniques. This method is more accurate than the other simulation-based approaches, while the computational complexity is much lower than that of the BDD-based approaches. Furthermore, it can be generalized to diagnose a sequential circuit with multiple design errors.

Based on the results of error diagnosis, designers can rectify the circuit under correction manually, or rely on a CAD tool to fix the bug(s) automatically. Automatic error correction has been pursued in several directions. The first direction is based on an *error model* that consists of a set of commonly encountered design errors (e.g., missing an inverter, or missing a wire [1]). Approaches along this direction try to match the errors with some error type(s) defined in the error model. Therefore, if the error is not covered by any modeled error type, this type of methods will fail to deliver a solution.

The second direction of error correction is to re-synthesize the function(s) of the identified potential error source(s) to fix the circuit [40,87,88,92,126]. The correction process involves three steps. First, compute the new function of the potential error signal(s). Second, synthesize the new function as a netlist in terms of the primary inputs. Third, replace the old function with the new function.

The third direction of error correction assumes that a certain degree of structural correspondence between the specification and the implementation will be provided by the designer or the synthesis tool [14,62]. A heuristic called *back-substitution* is then applied successively with an attempt to fix the implementation. The back-substitution is an operation that replaces a potential error signal in the implementation with its structural corresponding *good* signal in the golden reference specification. The details of this heuristic will be given in Chapter 8.

In Chapter 11, we discuss a hybrid method that combines the advantages of the second and the third types of approaches. For BDD-manageable circuits, we introduce an enhanced algorithm to perform correction incrementally. In this algorithm, the equivalence checking technique is applied to improve the overall correction quality. The correction quality is measured by the *recycling rate*, i.e., the ratio of the amount of logic that can be reused from the erroneous implementation to the amount of logic that needs to be newly added in the final correct implementation. A higher recycling rate means a higher error correction quality. For large circuits beyond BDD's capability, we propose an improved heuristic to apply the back-substitution technique by assuming that some structural correspondence data between the specification and the erroneous implementation are available.

1.2 Organization

The organization of this book is shown in Fig. 1.6. It is divided into two parts. The first part, Chapters 2 through 7, discusses formal equivalence checking. The second part, Chapters 8 through 11, discusses the algorithms for logic debugging including design error diagnosis and correction.

- In Chapter 2, we review the fundamental BDD-based finite state machine traversal techniques. Basic concepts of using BDD to represent a set of states and the machine's transition relation are first described. Speed-up techniques including early quantification and

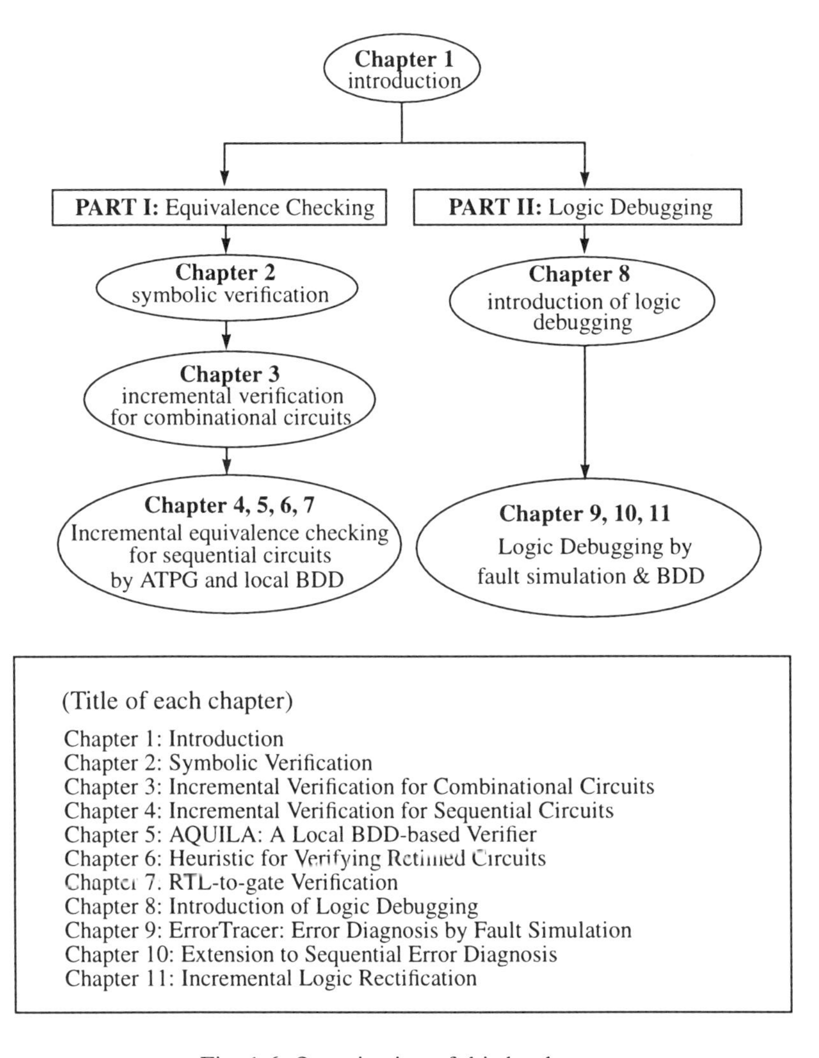

Fig. 1.6 Organization of this book.

reduction of image computation by cofactoring are then discussed.

- In Chapter 3, we review the state-of-the-art incremental algorithms for combinational verification including the substitution-based, learning-based, and transformation-based algorithms.

- In Chapter 4, we introduce an ATPG-based framework for checking sequential equivalence that is particularly suitable for a design without a known reset state. Several different definitions of sequential equivalence are discussed and compared. Also, speed-up techniques, e.g., an inductive algorithm for efficiently exploring the sequential similarity of the two circuits under verification, is introduced.
- In Chapter 5, we describe a BDD-based incremental verifier, AQUILA, for sequential circuits.
- In Chapter 6, we discuss a heuristic for verifying retimed circuits.
- In Chapter 7, we consider the problem of checking the equivalence for two designs with external don't care conditions. Also, we show how to integrate the pure finite state machine traversal with the incremental method.
- In Chapter 8, we review a number of representative methods for logic debugging. Algorithms using symbolic techniques as well as simulation techniques are described.
- In Chapter 9, we introduce a method of performing error diagnosis by a fault simulation process.
- In Chapter 10, we generalize the method in Chapter 9 for sequential circuits.
- In Chapter 11, we introduce an enhanced symbolic algorithm for automatic logic rectification.

The techniques discussed in several chapters of this book have been implemented as a logic verification and debugging system consisting of three programs: AQUILA, ErrorTracer, and AutoFix. These programs are described below:

- **AQUILA** is an equivalence verifier for large sequential circuits. This tool combines the advantages of incremental verification techniques with the traditional FSM traversal techniques. At the heart of this system is an inductive algorithm that extracts the sequential similarity to speed up the verification process. AQUILA has successfully handled many designs that are beyond the capability of the other traditional approaches. AQUILA is discussed in Chapter 5.

- **ErrorTracer** is a fault simulation-based approach to design error diagnosis. This approach is suitable for multiple design error diagnosis for combinational or sequential circuits. It is scalable and, thus, suitable for large designs. It has successfully diagnosed the entire set of ISCAS-85 benchmark circuits when injected with one or two random errors. Also, it is able to diagnose sequential circuits. ErrorTracer is discussed in Chapter 9.
- **AutoFix** is a hybrid tool for automatic design error correction in combinational circuits. This tool is very general in the sense that it does not rely on an error-model. Also, it can deal with circuits with multiple errors. Experimental results on a suite of industrial examples for the purpose of engineering change, and the entire set of ISCAS-85 benchmark circuits are given in Chapter 11 to demonstrate the effectiveness of this tool.

Fig. 1.7 shows a flow for using these programs for logic verification and debugging. Given two gate-level circuits, one is considered as the specifi-

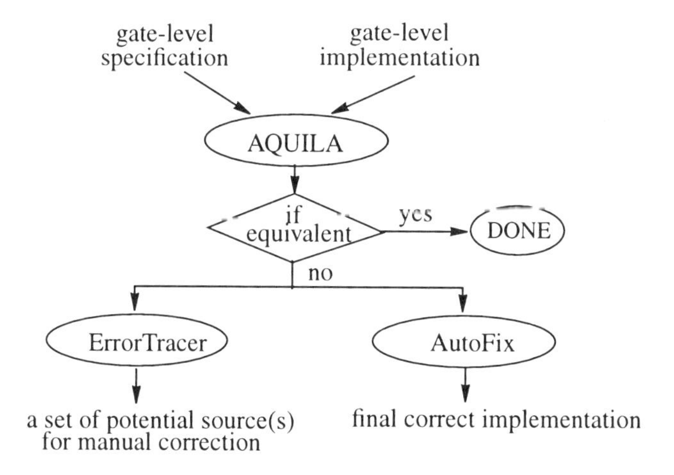

Fig. 1.7 System overview.

cation, and the other is considered as the implementation. We can first check their equivalence using AQUILA. If the given circuits are not equivalent, then two different debugging paths can be followed. For a heavily customized design, ErrorTracer is useful in generating a set of potential error source(s). The designer can manually fix the design based on this information. On the other hand, for the applications when automatic correction is desired, e.g., engineering change problem, AutoFix can be used to patch the implementation quickly. The binary codes of AQUILA and ErrorTracer for Sun Sparc workstation running SunOS4 or Solaris are available at ftp site *yellowstone.ece.ucsb.edu* through *anonymous* account or through the website: http://yellowstone.ece.ucsb.edu.

PART I

EQUIVALENCE CHECKING

Chapter 2

Symbolic Verification

In this chapter we discuss symbolic algorithms that rely on finite state machine (FSM) traversal to perform equivalence checking. FSM traversal is a process that explores the state space of a finite state machine using explicit or implicit state enumeration techniques. In the following discussion, we assume the reader is familiar with the basic concepts of the Binary Decision Diagram (BDD).

2.1 Symbolic Verification by FSM Traversal

As mentioned in Chapter 1, the first step in proving machine equivalence using symbolic techniques is to compute the reachable states of the product machine. This procedure is accomplished by finite state machine traversal. Given the state transition graph of a product machine, finite state machine traversal can be performed by traversing the state transition graph (STG) explicitly in either a depth-first, or a breadth-first manner. Relatively speaking, breadth-first traversal is more attractive because it allows an efficient computation using the BDD as will be shown later in this chapter. Fig. 2.1 illustrates the breadth-first strategy for FSM traversal.

In this iterative procedure, R_i represents the set of all reachable states at the i-th iteration. In the STG, R_i corresponds to the set of states that can be

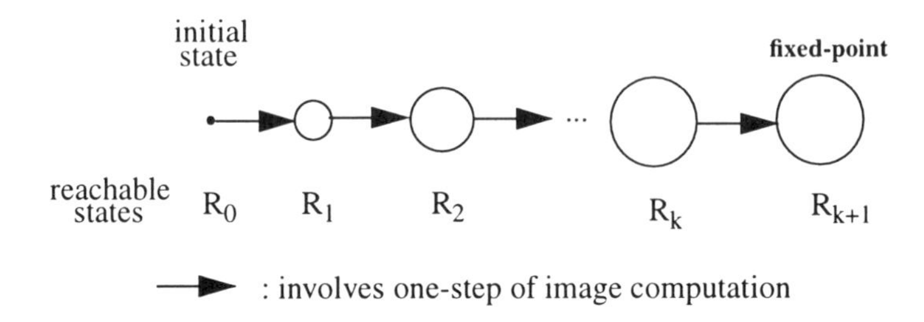

Fig. 2.1 The reachable states during the breadth-first FSM traversal.

reached in *i* or fewer transitions (clock cycles) from the reset states. Assume the reset state is s_0 and, therefore, $R_0 = \{s_0\}$. The set of reachable states grows monotonically at each iteration. This iterative procedure stops when it reaches a ***fixed-point***, meaning that the sets of the reachable states in two consecutive iterations are identical, i.e., $R_k = R_{k+1}$ for some positive integer *k*. At each iteration, the procedure involves the next-state computation to derive R_{i+1} from R_i. Next-state computation collects the next states reachable from any state in the current set of reachable states. If the set of reachable next states of R_i is denoted as N_i, then the set of reachable states at the iteration *i*+1, R_{i+1}, will be the union of R_i and N_i, i.e., $R_{i+1} = R_i + N_i$. The total number of iterations required equals the maximal depth (as explained below) of the STG plus 1.

Fig. 2.2 shows an example of explicit state traversal. Since we are only concerned about the transition behavior of the machine, the input/output information associated with each transition is omitted. The reachable states from s_0 at each iteration are listed in Table 2.1. In this example, the most distant state from s_0 is s_4, which takes 4 clock cycles to reach from the reset state s_0. Hence, this machine has a depth of 4, and the procedure performs 4 + 1 = 5 iterations of next-state computation to reach a fixed-point.

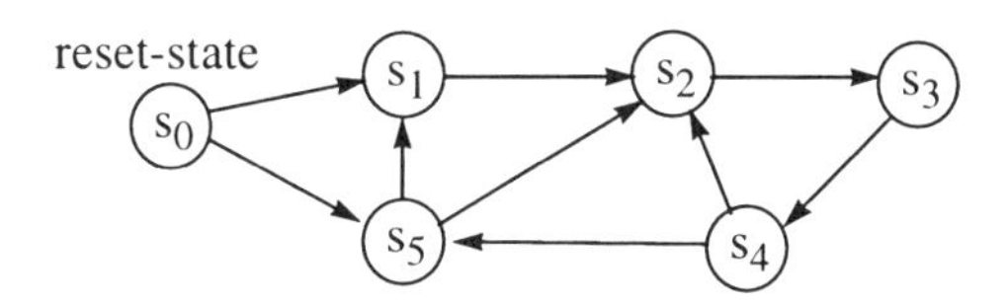

Fig. 2.2 An example for FSM traversal.

Table 2.1 Reachable states at each iteration of a breadth-first FSM traversal.

iteration	reachable states*
0	$\{s_0\}$
1	$\{s_0, \mathbf{s_1}, \mathbf{s_5}\}$
2	$\{s_0, s_1, \mathbf{s_2}, s_5\}$
3	$\{s_0, s_1, s_2, \mathbf{s_3}, s_5\}$
4	$\{s_0, s_1, s_2, s_3, \mathbf{s_4}, s_5\}$
5	$\{s_0, s_1, s_2, s_3, s_4, s_5\}$

* The states in bold face are the newly reached states in each iteration.

2.2 Implicit State Enumeration by BDD

Using symbolic BDD techniques, an FSM can be traversed without explicitly constructing the STG [21,39,42,124]. In these approaches, BDD is used to represent *a set of states* or *a machine's transition relation*. Based on these symbolic representations, the next-state computation can be performed efficiently and the state space is *implicitly* traversed. This technique is more powerful than the explicit graph-based methods and has successfully handled machines with more than 10^{20} states.

In the following, we describe implicit state enumeration by first reviewing the BDD representation of Boolean vectors, Boolean networks, and transition relations. Specifically, we will discuss:

- Representation of a *set of Boolean vectors* as a BDD.
- Representation of a *combinational Boolean network* as a BDD.
- Representation of a *transition relation of a machine* as a BDD.
- Next-state computation (image computation) based on a given transition relation represented as a BDD.

2.2.1 Set Representation

We will designate B for {0, 1} in the sequel. A BDD is simply a canonical form of a Boolean function, therefore, in order to represent a vector set as a BDD, we need to define a Boolean function that uniquely characterizes a given vector set.

Definition 2.1 [124] Let A be a set in the n-dimensional Boolean space B^n. The *characteristic function* of A is the function f_A: B^n -> B, defined by:

(1) $f_A(v) = 1$ if v is in A, where v is a Boolean vector in B^n.

(2) $f_A(v) = 0$ otherwise.

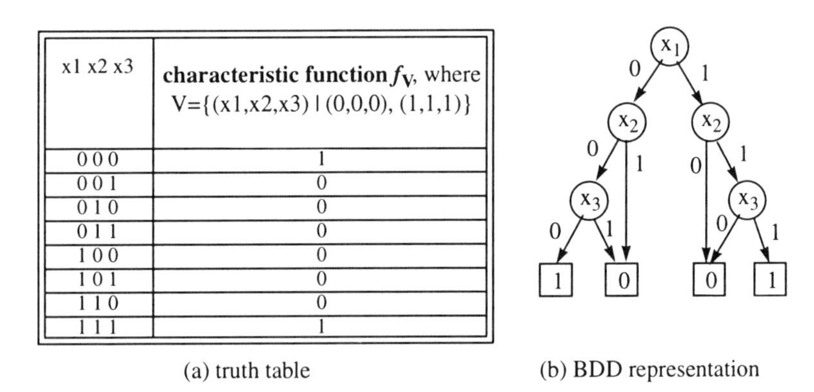

x1 x2 x3	**characteristic function** f_V, where V={(x1,x2,x3) | (0,0,0), (1,1,1)}
0 0 0	1
0 0 1	0
0 1 0	0
0 1 1	0
1 0 0	0
1 0 1	0
1 1 0	0
1 1 1	1

(a) truth table (b) BDD representation

Fig. 2.3 Characteristic function for a set of vectors.

Example 2.1 Consider an example in the Boolean space B^3. Let $A = \{(x_1, x_2, x_3) \mid (0,0,0), (1,1,1)\}$. The characteristic function of A, denoted as f_A: B^3–>B, maps (0,0,0) and (1,1,1) to '1', and all other vectors to '0'. The truth-table of this function is shown in Fig. 2.3. The Boolean function can be represented by a BDD.

2.2.2 Input/Output Relation

A Boolean network *C* with *m* primary inputs $X = \{x_1, x_2, ..., x_m\}$ and *n* primary outputs $Z = \{z_1, z_2, ..., z_n\}$ can be viewed as a set of *n* Boolean functions defined over B^m. For example, let *C* be $\{\lambda_1(X), \lambda_2(X),..., \lambda_n(X)\}$, where λ_i is an output function. To represent a Boolean network as a BDD, a characteristic input/output relation is defined to combine the *n* output functions together.

Given any input/output vector pair (v, w), if the output response of *C* is *w* when the input is *v*, then (v, w) is referred to as a *valid input/output combination*. A relation consisting of all valid input/output combinations uniquely defines a network's behavior. Hence, we can use the input/output relation to represent a Boolean network.

Definition 2.2 (*input/output relation*) Let *C* be a network from B^m to B^n. The *I/O relation* of *C* is a relation R_C: $B^m \times B^n$, and (v, w) is in R_C if $C(v) = w$, where *v* is an input vector, and *w* is an output vector.

Like a binary vector set, a relation can be converted to a characteristic function that maps every valid I/O combination to '1', and every invalid combination to '0'. However, unlike the primary output functions, this characteristic function is a function in terms of not only the primary input variables, X, but also the primary output variables, Z. Hereafter, we make no difference between a relation and its characteristic function. Computationally, the I/O relation can also be derived by the following formula:

$$R_C(x_1, \ldots, x_m | z_1, \ldots, z_n) = (z_1 \equiv \lambda_1(X)) \bullet (z_2 \equiv \lambda_2(X)) \ldots \bullet (z_n \equiv \lambda_n(X))$$

$$= \prod_{i=1}^{n} (z_i \equiv \lambda_i(X)) \qquad \text{where } (a \equiv b) \text{ corresponds to } (ab + \bar{a}\bar{b})$$

This formula defines a Boolean function that only evaluates to '1' when the i-th element of the given output vector, z_i, *agrees* with the i-th output response of the network, λ_i, with respect to the given input vector, for every i from 1 to n. Note that this definition of the characteristic function is the same as that derived for the Boolean satisfiability formulation of the test generation problem [24,82].

Example 2.2 Consider an example shown in Fig. 2.4(a). It has three primary inputs $\{x_1, x_2, x_3\}$ and two primary outputs $\{z_1, z_2\}$. Network C can be viewed as two Boolean functions:

- $z_1 = \lambda_1(x_1, x_2, x_3) = x_1 + x_2$
- $z_2 = \lambda_2(x_1, x_2, x_3) = (x_1 + x_2) \cdot x_3$

The I/O relation $R_C = (z_1 \equiv (x_1 + x_2)) \bullet (z_2 \equiv ((x_1 + x_2) \cdot x_3))$. The truth-table obtained by enumerating every valid I/O combination is shown in Fig. 2.4(b), and the BDD-representation of this I/O relation is shown in Fig. 2.4(c).

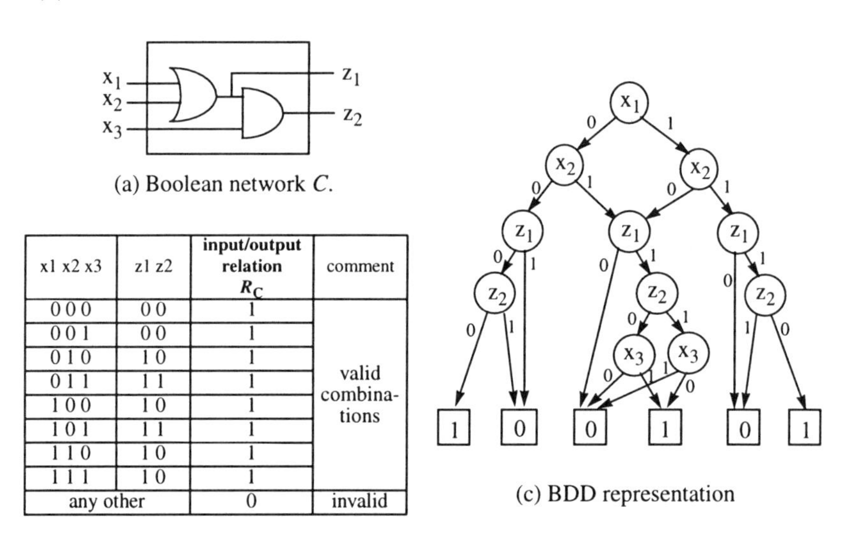

(a) Boolean network C.

x1 x2 x3	z1 z2	input/output relation R_C	comment
0 0 0	0 0	1	valid combinations
0 0 1	0 0	1	
0 1 0	1 0	1	
0 1 1	1 1	1	
1 0 0	1 0	1	
1 0 1	1 1	1	
1 1 0	1 0	1	
1 1 1	1 0	1	
any other		0	invalid

(c) BDD representation

(b) I/O relation of network C.

Fig. 2.4 Representing a Boolean network by its I/O relation.

It is worth mentioning that, although representing a network as an I/O relation is necessary for image computation (to be discussed later), its complexity is much higher than representing it as a set of output functions because the number of the supporting variables is $(m+n)$, instead of m. In many cases, construction of the BDD of the I/O relation may be impractical, even though BDDs of individual output functions can be constructed.

2.2.3 Transition Relation

The input format of a machine for symbolic FSM traversal could be a STG, or a Boolean network with flip-flops. If the given input is in STG form, then the first step is to perform state encoding which converts the STG into a representation in the Boolean domain.

As shown in Fig. 2.5, let M be an encoded FSM, denoted as a 6-tuple $M = (I, O, S, s_0, \delta, \lambda)$, where:

- I is the input space defined by input variables $\{x_1, x_2, ..., x_m\}$.
- O is the output space defined by output variables $\{z_1, z_2, ..., z_n\}$.
- S is the state space defined by state variables $\{y_1, y_2, ..., y_k\}$.
- s_0 is the known reset state.
- δ is a set of transition functions.
- λ is a set of output functions.

Definition 2.3 (*transition relation*) Consider a machine $M = (I, O, S, s_0, \delta, \lambda)$. The *transition relation* of M is a relation denoted as T: $B^m \times B^k \times B^k$, where m and k are the dimensions of the input space and the state space, and (v, p, q) is in T if machine M will transition from state p to state q under the input vector v.

In the Huffman model shown in Fig. 2.5, we denote the next-state lines (inputs of the flip-flops) of M as $\{t_1, t_2, ..., t_k\}$. The transition relation of a machine is simply the input/output relation of the machine's combinational portion, ignoring the output functions. Hence, it can be computed by the following formula:

$$T(x_1, \ldots, x_m | y_1, \ldots, y_k | t_1, \ldots t_2, t_k) = (t_1 \equiv \delta_1) \bullet (t_2 \equiv \delta_2) \ldots \bullet (t_k \equiv \delta_k)$$

$$= \prod_{i=1}^{k} (t_i \equiv \delta_i(X, Y))$$

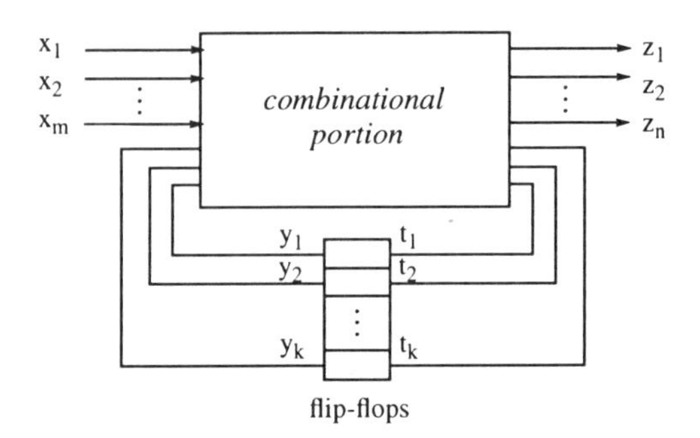

Fig. 2.5 Notation of a machine's Huffman model.

The transition relation uniquely determines the sequence of states the machine will go through in response to any input sequence. However, in the process of next-state computation, the specific input vector to exercise a particular transition is not of interest. We are only interested in knowing *if there exists a transition that brings the machine from state p to state q*. In other words, we only need a relation that captures the *connectivity* of the STG.

Definition 2.4 (*smoothed transition relation*) Let $M = (I, O, S, s_0, \delta, \lambda)$. The *smoothed transition relation* of M is a relation denoted as $T_{smoothed}$: $B^k \times B^k$, where k is the dimension of the state space, and (p, q) is in T if *there exists* a transition from state p to state q.

The smoothed transition relation is computed by *smoothing out* every primary input variable from the transition relation, where smoothing a variable is to *existentially quantify* a variable defined as follows.

Definition 2.5 (*existential quantification*) Let $f(x_1, x_2, ..., x_m)$ be a Boolean function. The existential quantification of f with respect to the variable x_i is:

$$(\exists x_1)f = f_{x_i} + f_{\bar{x}_i}$$

where $f_{x_i} = (x_1, \ldots, x_i = 1, \ldots, x_m)$, and $f_{\bar{x}_i} = (x_1, \ldots, x_i = 0, \ldots, x_m)$.

Existential quantification for a set of variables, e.g., $X = \{x_1, x_2, ..., x_m\}$, is defined as a sequence of single-variable smoothing operations:

$$(\exists X)f = \exists x_1(\exists x_2...(\exists x_m f))$$

It can be shown that the order of the variables is not relevant. Plugging the smoothing operations into the transition relation, we will have a smoothed transition relation in terms of the present-state line variables, $\{y_1, y_2, ..., y_k\}$, and next-state line variables, $\{t_1, t_2, ..., t_k\}$, only:

$$T_{smoothed}(y_1, ..., y_k | t_1, ...t_2, t_k) = (\exists x_1 x_2...x_m)((t_1 \equiv \delta_1) \bullet (t_2 \equiv \delta_2)... \bullet (t_k \equiv \delta_k))$$
$$= (\exists x_1 x_2...x_m) \prod_{i=1}^{k} (t_i \equiv \delta_i(X, Y))$$

2.2.4 Next-State Computation

Smoothed transition relation defines a projection from the present state space to the next state space as shown in Fig. 2.6. Since a state could reach

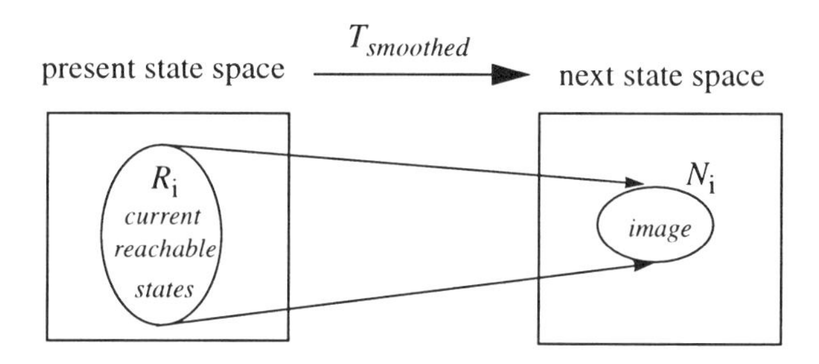

Fig. 2.6 Next state computation using smoothed transition relation.

more states than one, and multiple states can reach the same next state, $T_{smoothed}$ is a many-to-many projection. Based on this projection, the set of reachable next states, N_i, is the image of the current reachable set, R_i.

Definition 2.6 (*image*) Let T be a projection from domain B^m to co-domain B^n, denoted as T: $B^m x B^n$, and A be a set of Boolean vector in B^m. The *image* of A is a set of Boolean vectors in B^n, defined by:

$$Image(T, A) = \{w \in B^m | (v, w) \in T, v \in A\}$$

The next states reachable from the current reachable set R_i is therefore the image of R_i under the smoothed transition relation. In terms of Boolean operations, the next-state computation can be performed as:

$$\begin{aligned} \textit{reachable next states } N_i &= Image(T_{smoothed}, R) \\ &= (\exists y_1 y_2 \ldots y_k)(R_i \bullet T_{smoothed}) \\ &= (\exists y_1 y_2 \ldots y_k)\left(R_i \bullet \left((\exists x_1 x_2 \ldots x_m) \prod_{i=1}^{k} (t_i \equiv \delta_i(X, Y))\right)\right) \end{aligned}$$

Where $(R_i \bullet T_{smoothed})$ can be interpreted as the set of present-next state pair (p, q). This product evaluates to '1' if and only if the present-next state pair (p, q) is valid under the transition relation $T_{smoothed}$ and the present state p is in the current reachable set R_i. Further smoothing out the present state variables, we arrive at a set of next-states which is the set of reachable next states N_i.

2.2.5 Complete Flow

Fig. 2.7 shows the complete flow of the BDD-based symbolic verification. It primarily consists of three stages:

(1) Pre-processing stage: involves the state encoding if the given specification is not encoded yet, and the construction of the product machine.

(2) Reachable states computation: performs the FSM traversal using symbolic techniques as described in the previous subsections.

(3) Tautology checking: checks if every primary output of the product machine is a tautology '0' for any input vector and reachable state.

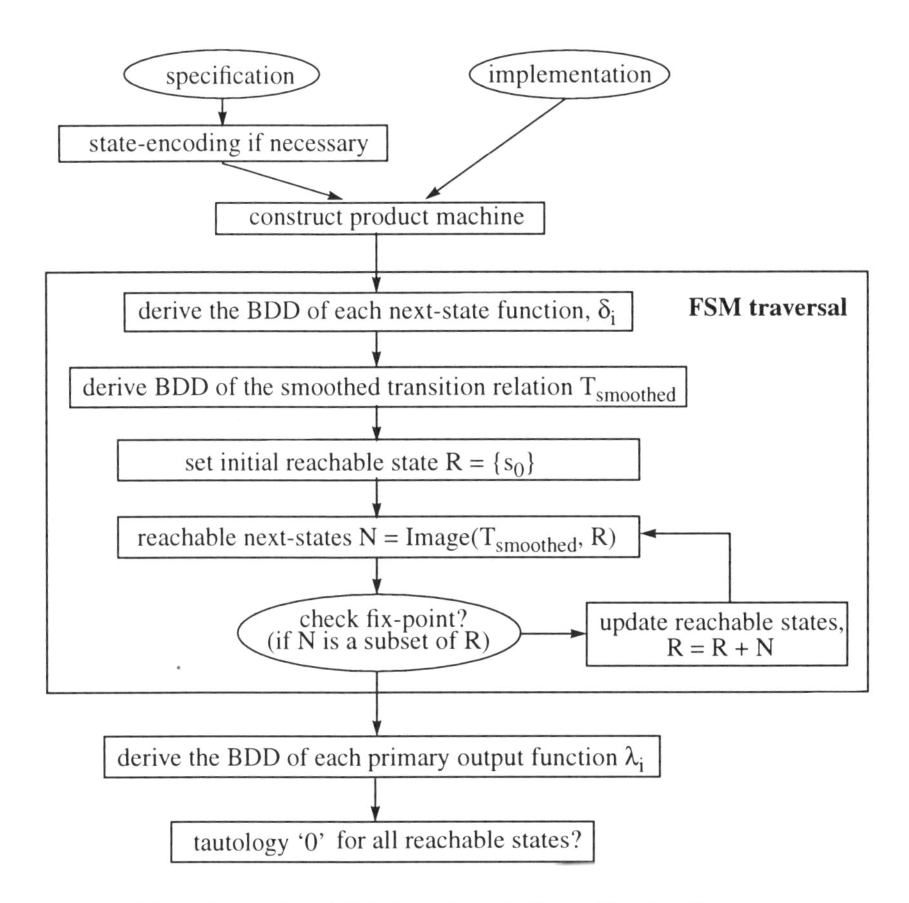

Fig. 2.7 Primitive BDD-based symbolic verification flow.

The tautology check can be translated into a Boolean predicate, that is:

$$\underbrace{(\exists x_1 x_2 \ldots x_m (\textstyle\sum_{i=0}^{n} \lambda_i))}_{\text{distinguishable states}} \bullet R = \phi$$

A *distinguishing state*, *s*, is a state in which the product machine produces '1' at some primary output λ_i for some input vector *v*, i.e., $\lambda_i(v, s) = 1$. The above predicate is true when there does not exist a *distinguishing state* that is *reachable*. This predicate can be determined by constructing the BDD of the left-hand-side and check if it is a zero-function BDD. This predicate is true if and only if the two machines are functionally equivalent.

2.2.6 Error Trace Generation

When two CUVs are not equivalent, it is usually desirable to derive an input sequence that distinguishes the two CUVs. The process of generating such a sequence, often referred to as *error trace generation*, consists of a series of *pre-image computations*.

Definition 2.7 (*pre-image*) Let T be a projection from domain B^m to co-domain B^n, denoted as T: $B^m x B^n$, and A be a set of Boolean vector in B^n. The *pre-image* of A is a set of Boolean vectors in B^m, defined by:

$$pre\text{-}image(T, A) = \{p \in B^m | (p, r) \in T, r \in A\}$$

The pre-image computation is similar to the image computation described earlier. Assume that the co-domain variables are $\{z_1, z_2, ..., z_n\}$. Then the pre-image can be computed by the following formula:

$$pre\text{-}image(T, A) = (\exists z_1 z_2 \ldots z_n)(T \bullet A)$$

In the step of tautology check, if a set of reachable distinguishing state, E, is found, then the error trace generation is a *sequential justification problem* which finds an input sequence that brings the product machine from the reset state to a state in the set of the reachable distinguishing states, E. Theoretically, finding *every* error trace can be done symbolically. However, the complexity is often too high and thus infeasible. Here, we only discuss the procedure of generating one error trace. Suppose the set of distinguishing states E intersects with R_k, but not R_i for i from 0 to k-1, as shown in Fig. 2.8. It can be shown that the minimum length of any error trace is $k+1$.

Fig. 2.9 shows an algorithm for finding a minimum length error trace. The inputs to this algorithm are the transition relation T, the primary output

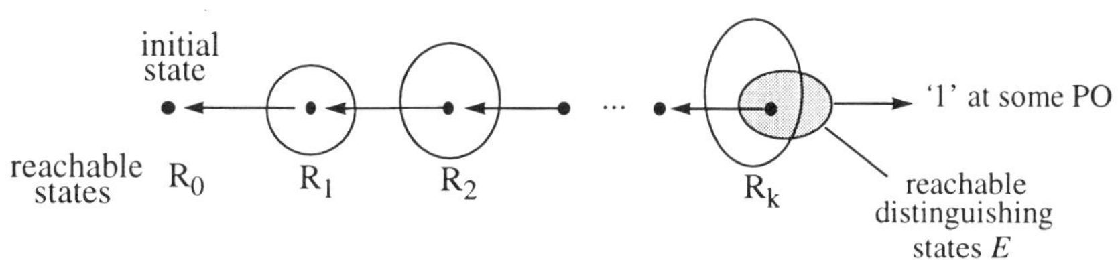

Fig. 2.8 An Error Trace.

functions, λ, and the reachable states R[i], i = 0, 1, ..., k, of the first (k+1) iterations. Also, we use the array v to store the (k+1) vectors of the error trace, and the array s to store the states of the product machine in response to the error trace.

```
Error_trace_generation(T, λ, R)
    T: transition relation of the product machine.
    λ: primary output functions of the product machine.
    R: reachable states at each iteration of FSM traversal.
{
    P: a set of vectors in terms of B^m x B^k.
    v: a distinguishing sequence to be generated.
    s: a sequence of states to which the error trace will bring the machine.

    /*----- step 1: determine the last vector -----*/
    (v[k] | s[k]) = pick-one-minterm-from ( (Σ_{i=0}^{n} λ[i] • R[k]) ).

    /*----- step 2: determine the first k vectors -----*/
    for(i=k-1; i>=0; i--){
        P = Pre-image( T, { s[i+1] } );
        (v[i] | s[i]) = pick-one-minterm-from( (P • R[i]) ).
    }
    return(v);
}
```

Fig. 2.9 Error trace generation algorithm.

Step 1 generates the last vector of the error trace. Again, the formula, $(\sum_{i=0}^{n} \lambda[i] \bullet R[k])$, characterizes the combinations of input vectors and states in R[k] that can produce '1' at *some* primary output. Only one such combination is selected and denoted as (v[k] | s[k]).

Step 2 is a k-iteration loop that generates the first k input vectors. Each iteration identifies the combination of an input vector v and a state s that satisfies:

- (v | s) as applied to the combinational portion of the product machine leads to the state s[k+1], and
- s is in R[i].

Combinations satisfying this criterion are identified by pre-image computation as shown in the algorithm. If there are many such combinations, then only one will be selected and denoted as (v[i] | s[i]).

2.3 Speed-up Techniques

In the entire verification algorithm, building the BDDs of the transition relation and performing the image-computation, are the two steps that may cause memory explosion. A variety of techniques have been proposed to address this issue. In the following, we discuss three techniques: (1) an efficient construction method for the smoothed transition relation, (2) a reduction technique for the image computation, and (3) a reduction technique for the overall reachable state computation.

2.3.1 An Efficient Method for Constructing Transition Relation

Touati et. al. [124] introduced a technique to reduce the complexity of building the smoothed transition relation. Recall that the smoothed transition relation T_{smoothed} is expressed as:

$$T_{smoothed}(y_1, \ldots, y_k | t_1, \ldots t_2, t_k) = (\exists x_1 x_2 \ldots x_m)((t_1 \equiv \delta_1) \bullet (t_2 \equiv \delta_2) \ldots \bullet (t_k \equiv \delta_k))$$

Building this relation requires k-1 Boolean AND operations and m existential quantifications. The k-1 Boolean AND operations can be naively performed from left to right sequentially as shown in Fig. 2.10. Here, an atomic formula $(t_i \equiv \delta_i)$ for some integer i between 1 and k is called a *sub-relation*, and the conjunction of a number of sub-relations is called a *partial product*:

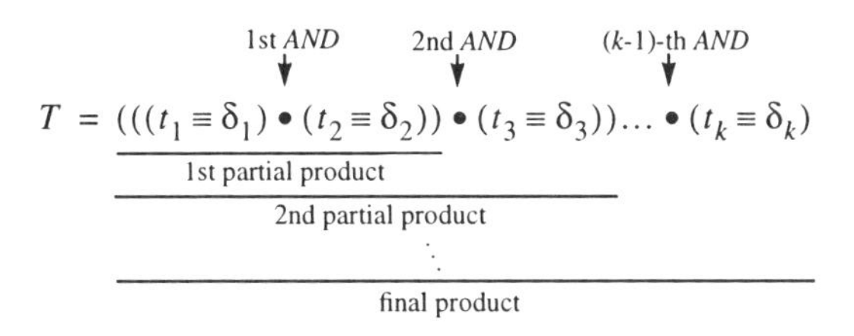

Fig. 2.10 Building transition relation from left-to-right sequentially.

The efficiency can be improved by decomposing the process into a balanced binary tree as shown in Fig. 2.11. The tree is constructed in a bottom-up manner. Each leaf node corresponds to a sub-relation, and the root node corresponds to the final transition relation. This process is more efficient because the partial products associated with each node are relatively simpler than those in the above sequential method.

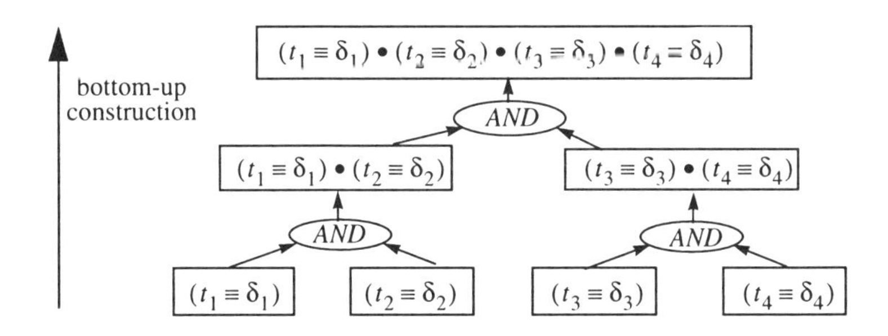

Fig. 2.11 Building transition relations by a balanced binary tree.

Once a transition relation is derived, the primary input variables are smoothed out to obtain the smoothed transition relation, $T_{smothed}$. However, for most cases, T_{smothed} can be constructed more efficiently without building the exact transition relation. A technique called *early quantification* that existentially quantifies the primary input variables as early as possible has been proposed [124]. The following example illustrates this idea.

Example 2.3 Given a machine with two next-state line functions: $\delta_1(x_1, x_2, y_1, y_2)$ and $\delta_2(x_2, x_3, y_1, y_2)$. By analyzing the supporting variables of these two functions, it can be found that x_2 is the only input variable shared by both sub-relations. Therefore, we can rephrase the smoothed transition relation as follows:

$$T_{smoothed} = (\exists x_2)(\underbrace{(\exists x_1)(t_1 \equiv \delta_1(x_1, x_2))}_{\textit{early-smoothed } \boldsymbol{T_1}} \bullet \underbrace{(\exists x_3)(t_1 \equiv \delta_1(x_2, x_3))}_{\textit{early-smoothed } \boldsymbol{T_2}})$$

In this new formula, some smoothing operations are distributed into sub-relations and are performed as early as possible, e.g., T_1 is early quantified by x_1, and T_2 is early quantified by x_3, respectively. This early quantification is appropriate because T_1 is the only sub-relation depending on x_1, and T_2 is the only sub-relation depending on x_3. Because a smoothing operation always reduces the complexity of a Boolean formula in terms of its BDD size, the above formula requires less memory and is more efficient than the original one, which requires building every sub-relation exactly and performing (k-1) Boolean *AND* operations before performing the smoothing operations. The next example shows the generalization of this idea to the construction method based on binary trees.

Example 2.4 Given a machine with four next-state line functions: $\delta_1(x_1, x_2, y_2)$, $\delta_2(x_2, x_3, y_3)$, $\delta_3(x_3, x_4, y_4)$, and $\delta_4(x_4, x_1, y_1)$. The Boolean operations to be performed in the binary tree are shown in Fig. 2.12.

For each partial product constructed during the bottom-up process, a supporting variable is smoothed out immediately if it is not a supporting variable of *any other* partial product at the same level. The final outcome of this tree is the smoothed transition relation, instead of the original transition relation.

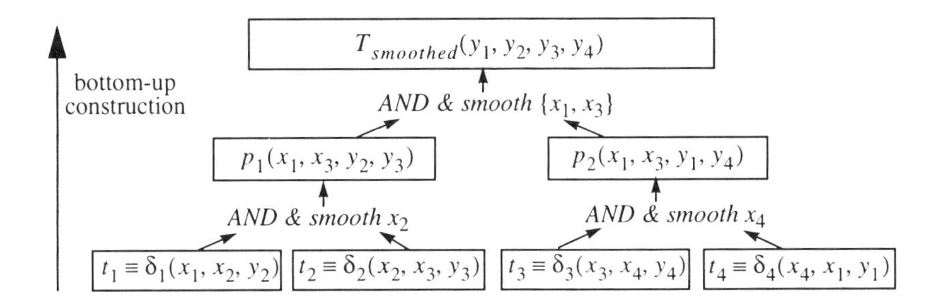

Fig. 2.12 An example for early quantification.

2.3.2 Reduction on Image Computation

Even with the above tree-based construction method and early quantification technique, the exact smoothed transition relation is still hard to build sometimes. Another powerful speed-up technique [43] is to perform image computation without building the exact transition relation.

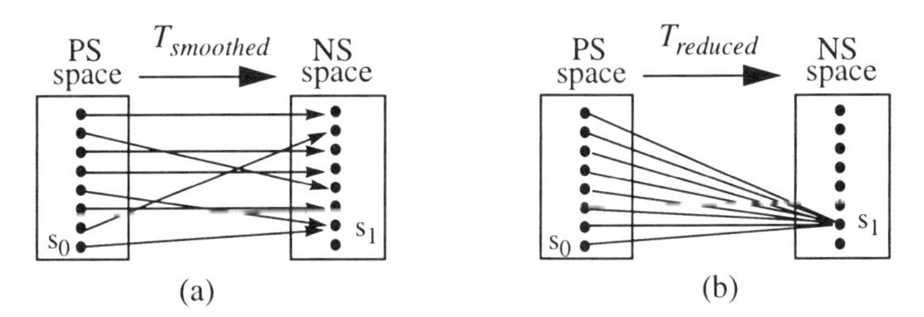

Two relations are *consistent* for the image of $\{s_0\}$

Fig. 2.13 Reduced transition relation for efficient image computation.

Consider the example in Fig. 2.13 where we wish to compute the image of the state $\{s_0\}$. Each dot in the figure represents a state. An arc from a state p in PS space to a state q in NS space means that there exists a transition from state p to state q. Fig. 2.13(a) shows the projection defined by the exact smoothed transition relation and Fig. 2.13(b) shows another projection, called *reduced transition relation* T_{reduced}.

It can be seen that these two transition relations are *consistent* for the transition originating from the state $\{s_0\}$. They both lead to the same result $\{s_1\}$ if used for image computation. In terms of the BDD size, T_{reduced} is potentially smaller than T_{smoothed} because it has a simpler projection pattern, i.e., it projects every present state to the same next state $\{s_1\}$. This example provides an important observation: *for a particular set of present states, many projections other than the exact transition relation can be used to compute the set of next states without losing accuracy.* In the following, we discuss a concept called generalized cofactor used to speed up the image computation based on this observation.

Definition 2.8 (*generalized cofactor* [42,124]) Let *f*: $B^m x B^n$->B, and *c*: B^m->B be Boolean functions. The generalized cofactor f_c: $B^m x B^n$->{0, 1, *d*}, where *d* represents the don't care logic value, is an incompletely specified Boolean function defined by:

- $f_c(v, w) = f(v, w)$ if v is in c.
- $f_c(v, w) = d$ if v is not in c.

As in Fig. 2.13, if *f* is viewed as a relation or projection from domain B^m to co-domain B^n, and *c* is a set of vectors in B^m, then any projection that agrees with *f* on *c* is a cofactor of *f*. It is safe to replace *f* with any cofactor for computing the image of *c*. To determine the best cofactor of minimal BDD size for a given function *f* and its constraint *c* is an optimization problem. Fig. 2.14 shows an algorithm [124] that can be performed efficiently by a single bottom-up traversal of BDDs of *f* and *c*. More advanced algorithms can be found in [27,58,120].

During the FSM traversal, suppose *R* is the current set of reachable states. Then, a cofactor of the smoothed transition relation, denoted as T_{reduced}, can be efficiently computed by the conjunction of every cofactored sub-relation:

$$T_{reduced} = \left(\prod_{i}^{k} (t_i \equiv \delta_i)\right)_R = \prod_{i}^{k} (t_i \equiv \delta_i)_R$$

Cofactor(f, c)
{
  if(c == 1 or f is a constant-BDD) return(f);
  if(c_{x1} == 0) return($cofactor(f_{\overline{x1}}, c_{\overline{x1}})$);
  if($c_{\overline{x1}}$ == 0) return($cofactor(f_{x1}, c_{x1})$);
  return($x_1 \cdot cofactor(f_{x1}, c_{x1}) + \overline{x_1} \cdot cofactor(f_{\overline{x1}}, c_{\overline{x1}})$);
}

Fig. 2.14 General cofactor algorithm.

In other words, the cofactoring operation can be distributed into each sub-relation. Applying this formula, along with the early quantification technique, to the tree-based construction method, one can replace every leaf node with its cofactored version as shown in Fig. 2.15.

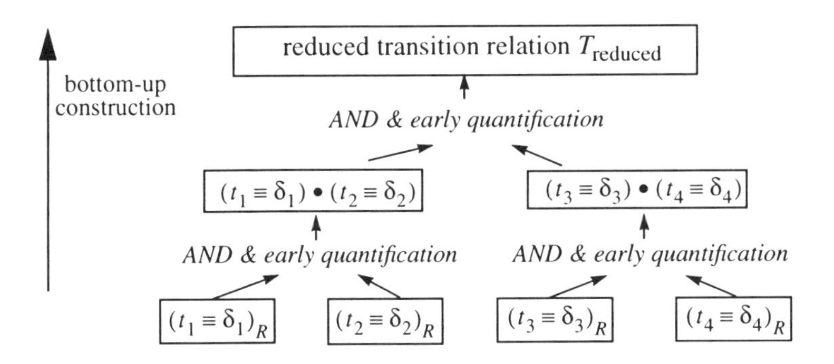

Fig. 2.15 Building the reduced transition relation.

We must point out that the reduced transition relation is dependent on the current set of reachable states. Therefore, it needs to be re-computed at each iteration of the FSM traversal process. However, this overhead is usually paid for by the significantly reduced complexity of the image computation.

2.3.3 Reduction on Reachable State Computation

In general, the image of a smaller set of states is easier to compute than that for a larger set of states because more flexibility is available in minimizing the reduced transition relation. Based on this observation, the reachable state computation can be further improved.

Recall that the set of reachable states at iteration i+1 is computed by:

$$N = Image(T_{smoothed}, R_i)$$
$$R_{i+1} = R_i + N$$

where N is the set of reachable next-states of R_i.

Suppose R_{i-1} is also available in addition to R_i while computing R_{i+1}. Let R_i be decomposed into two subsets, R_{i-1} and $(R_i - R_{i-1})$. Then the set of reachable next-states of R_i is the union of the images of these two sets under the smoothed transition, i.e.,

$$N = Image(T_{smoothed}, R_{i-1}) + Image(T_{smoothed}, (R_i - R_{i-1}))$$

It follows that,

$$R_{i+1} = (R_i + N) = R_i + Image(T_{smoothed}, R_{i-1}) + Image(T_{smoothed}, (R_i - R_{i-1}))$$

Because the second term $Image(T_{smoothed}, R_{i-1})$ is a subset of the first term R_i, it can be eliminated, and resulting in the following formula:

$$R_{i+1} = R_i + Image(T_{smoothed}, (R_i - R_{i-1}))$$

This new formula is relatively more efficient than the original one because the set of states involved in the image computation, $(R_i - R_{i-1})$, is smaller than the original set, R_i. As a result, it allows greater flexibility in reducing the computational complexity using the concept of cofactoring. From another viewpoint, this formula is more efficient because it avoids the unnecessary computation of $Image(T_{smoothed}, R_{i-1})$ while deriving R_{i+1}.

2.4 Summary

In this chapter, we have discussed the symbolic verification algorithms that rely on the finite state machine traversal using BDD techniques. The key issues discussed include modeling a set of states and a machine's transition relation as a BDD, and the computation of the reachable states through a sequence of symbolic image computations. Unlike the STG-based methods that enumerates the states of a machine explicitly, these approaches perform implicit state enumeration without building the state transition graph. In addition to the primitive FSM traversal algorithm, we also discussed several speed-up techniques, including a tree-based construction method for the transition relation, early quantification, and a reduction technique based on cofactoring.

Chapter 3

Incremental Verification for Combinational Circuits

In this chapter we discuss the incremental algorithms that explore the structural similarity between two circuits to speed up the verification process. Three types of algorithms, namely substitution-based, learning-based, and transformation-based algorithms, are described and compared.

3.1 Substitution-Based Algorithms

During the design cycle, equivalence checking is often performed on two circuits with significant structural similarity. For instance, checking if a timing-optimized design is functionally equivalent to its original version, or comparing a transistor-level netlist with its gate-level model. For such applications, Brand's incremental verification algorithm [13] can verify very large designs.

3.1.1 Brand's Algorithm Using ATPG

In the following we assume that both circuits have single output. The discussion can be easily generalized to multiple output circuits. As mentioned earlier, proving the equivalence of two circuits is reduced to proving that no input vector produces '1' at the miter's output signal g. Instead of building the BDD-representation of the miter like the symbolic approaches, equivalence checking can also be formulated as a search problem that searches for a *distinguishing vector*, to which the two circuits under verification produce different output responses. If no distinguishing vector is found after the entire search space is exhausted, then the two circuits are proven equivalent. Otherwise, a counter-example is generated to disprove the equivalence. Since a distinguishing vector is also a test vector for miter's output g stuck-at-0 fault, equivalence checking is reduced to a test generation process for g stuck-at-0 fault. However, directly applying ATPG to check the output equivalence could be very time-consuming for a large design. It is known [13] that the complexity can be reduced dramatically by exploring the structural similarity between the two CUVs.

Definition 3.1 (*signal pair*): Let a_1 and a_2 be internal signals. (a_1, a_2) is called a signal pair if a_1 and a_2 are from different circuits, e.g., a_1 is from C_1 and a_2 is from C_2, or vice versa.

Definition 3.2 (*equivalent pair*): (a_1, a_2) is called an equivalent (signal) pair if the binary values of signal a_1 and a_2 in response to any input vector are identical.

Definition 3.3 (*permissible pair*) [100]: (a_1, a_2) is called a permissible (signal) pair if a_1 is a *permissible function* of a_2, i.e., replacing signal a_1 by a_2 in the miter does not change the functionality of the miter's output signal g. Note that an equivalent pair is a permissible pair, but not vice versa.

Roughly speaking, two circuits are considered as *similar* if large percentage of signals in the circuits belong to equivalent pairs, or permissible pairs. The overall verification process [13] utilizing these internal equivalent and permissible pairs is shown in Fig. 3.1. It consists of three stages: (1) pairing up candidate pairs, (2) incrementally pruning the miter, and (3) checking the equivalence of each primary output pair.

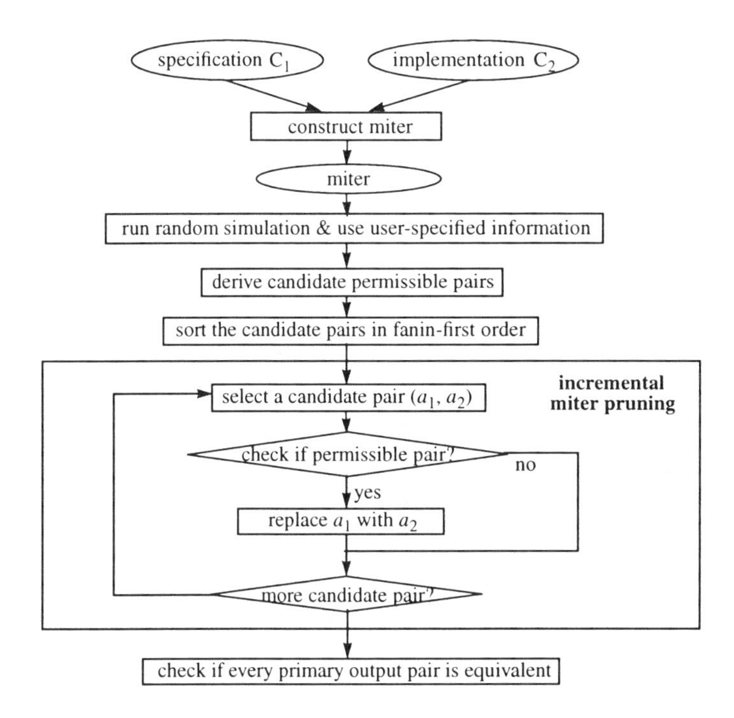

Fig. 3.1 Incremental verification for combinational circuits.

Pairing up candidate pairs

After the miter is built, the candidate equivalent or permissible pairs are first derived by random simulation, naming information, and/or user-specified information. After simulating a large number of random vectors, a signal pair (a_1, a_2) is regarded as a candidate equivalent pair if a_1 and a_2 have the same value in response to every input vector applied. For combinational circuits, random simulation often yields a satisfactory result in pairing up candidate equivalent pairs [72]. However, the candidate permissible pairs that are not equivalent pairs cannot be identified in this way. The candidate permissible pairs can be derived by either paring up signal pairs of the same signal names or using external information specified by the user. If neither the naming nor the user-specified information is available, then every signal pair could be regarded as a candidate permissible pair optimistically to start with. In that case, the total number of candidate pairs would be $(n_1 \cdot n_2)$, where n_1 and n_2 are the numbers of signals in C_1 and C_2, respectively. This number is usually too large, and could make the subsequent process (which iterates through every candidate pair) much too expensive. An alternative is to focus on identifying candidate equivalent pairs only. Evidence shows that many circuits can be verified efficiently by identifying equivalent signal pairs only [62,75,97].

Pruning the miter incrementally

Once the candidate pairs are derived, a sequence of operations that reduces the size of the miter incrementally is followed. First, the candidate pairs are sorted in a *fanin-first* order (i.e., a signal is placed after its transitive fanin signals) based on the circuit's topology. After the sorting, the algorithm enters a loop in which each iteration selects a candidate pair, and checks if the target pair is permissible. Checking permissibility is the most important step in this algorithm that can be performed by running ATPG using the model shown in Fig. 3.2 (a). In this model, the two signals

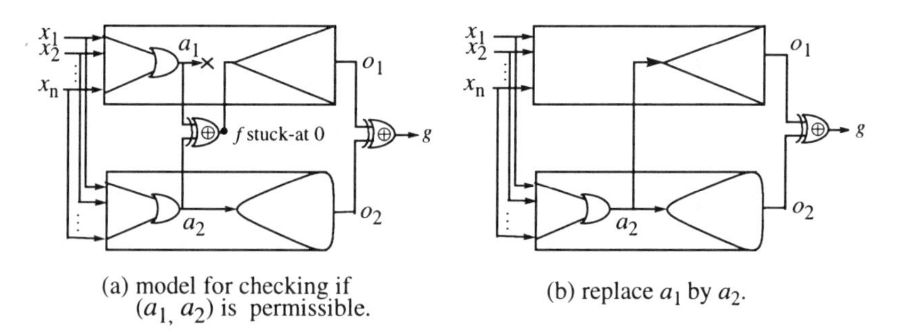

Fig. 3.2 Pruning the miter by finding a permissible pair.

in the target pair are connected to an exclusive-OR gate. The output of this gate, denoted as *f*, further replaces the first signal a_1. It will be shown later that if signal f stuck-at-0 fault is untestable, then it is safe to replace a_1 by signal a_2 without changing the functions of o_1 and the miter's output g. If a candidate pair (a_1, a_2) is found indeed permissible, the first signal a_1 is replaced by the second signal a_2 to simplify the miter as shown in Fig. 3.2(b). During incremental verification, the miter is simplified step-by-step from the primary input side towards the primary output side, so that the verification is done in stages and the computational complexity is dramatically reduced. Similar procedures for pruning the circuit have been used for test generation by Boolean satisfiability and energy minimization methods [25,26,82].

Definition 3.4 (*merge point*) In what follows, replacing a signal a_1 in C_1 by a signal a_2 in C_2 is also called a *merge operation*, and the signal a_2 is called a *merge point*.

Lemma 3.1 [13] Signal pair (a_1, a_2) is permissible (i.e., a_1 is a permissible function of a_2) if and only if signal f *stuck-at*-0 fault is untestable in the model shown in Fig. 3.2.

Explanation: By definition, signal a_1 is a permissible function of a_2 iff there exists no input vector that can produce (0, 1) or (1, 0) at (a_1, a_2), and make this difference observable at the miter's output g. This is exactly the condition of *activating* signal f stuck-at-0 fault and *propagating* the fault effect to the miter's output g. Therefore, (a_1, a_2) is permissible if and only if f stuck-at-0 is untestable.

Identifying *inverse permissible pairs* is also very important in many cases, e.g., verifying a circuit after technology mapping against its original version. A signal pair (a_1, a_2) is called an inverse permissible pair if a_1 can be replaced by the complement of signal a_2 without changing the function of the miter's output. The model for checking an inverse permissible pair and its corresponding transformation is shown in Fig. 3.3. The only difference between this model and the one for checking a permissible pair is the addition of an inverter.

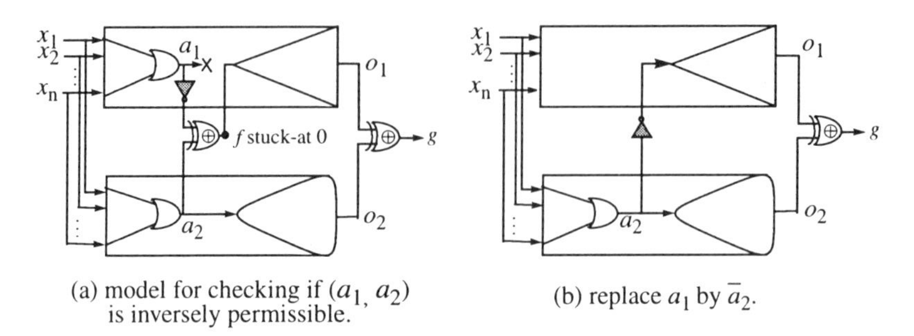

Fig. 3.3 Pruning the miter by finding an inverse permissible pair.

Checking the equivalence of primary outputs

After the miter is simplified by merging all internal permissible pairs, the equivalence of each primary output pair is checked by running ATPG for g stuck-at-0 fault. If a test vector v is found, then v serves as a counter-example to prove that two CUVs are not equivalent. On the other hand, if ATPG concludes that g stuck-at-0 fault is untestable, then the two CUVs are functionally equivalent.

3.1.2 Enhancement by Local BDD

The ATPG-based approach is efficient if the two given circuits have significant structural similarity. But for circuits optimized through extensive transformations where only a small percentage of signals have equivalent counterparts, the pure ATPG-based technique discussed above may not have satisfactory performance. For such cases, the process of proving equivalent internal signal pairs can be enhanced by the use of *local BDD's*. A local BDD refers to a BDD whose supporting variables could be internal signals, instead of the primary inputs. In the following, we discuss an algorithm [62]. Similar ideas were also proposed by Jain et al. [71], Kuehlmann and Krohm [75], and Matsunaga [97].

A sufficient condition of functional equivalence

We first define the discrepancy function, and then show a sufficient condition for the equivalence of a signal pair.

Definition 3.5 (*Discrepancy function*): An input vector v is a distinguishing vector for a signal pair (a_1, a_2) if the application of v can produce (0, 1) or (1, 0) at a_1 and a_2. The characteristic function of the set of distinguishing vectors is called the *discrepancy function* and denoted as $Disc(a_1, a_2)$. If the functions of a_1 and a_2 are denoted as $F(a_1)$ and $F(a_2)$, respectively, then $Disc(a_1, a_2) = F(a_1) \oplus F(a_2)$.

By definition, a signal pair (a_1, a_2) is equivalent if and only if its discrepancy function $Disc(a_1, a_2)$ is tautology '0'. One way to prove the equivalence of a signal pair would be to construct the global[1] BDD representation of their discrepancy function. However, for larger designs, this may not be feasible due to the explosion of the BDD size. In many cases, the equivalence of a signal pair can be claimed by only building a local BDD representation of the discrepancy function. This is based on the sufficient condition that if there exist a cutset λ in the input cones of a_1 and a_2 such that no value combination at the cutset can produce (0, 1) or (1, 0) at (a_1, a_2), then (a_1, a_2) is an equivalent pair. We denote the discrepancy function with respect to a cutset λ for signal pair (a_1, a_2) as $Disc_\lambda(a_1, a_2)$.

Property 3.1 Signals a_1 and a_2 are equivalent if there exists a cutset λ in the input cones of a_1 and a_2 such that $Disc_\lambda(a_1, a_2)$ is tautology zero.

The false negative problem

The above property is only a sufficient condition because the signals in the selected cutset might be *correlated* (*or not independent*). A cutset λ is correlated if there exist impossible value combination(s) at the cutset, over all possible vectors at the primary inputs. Since it is only a sufficient condition, the *false negative problem* may arise. That is, the target pair is indeed equivalent, but with respect to the selected cutset λ, $Disc_\lambda(a_1, a_2)$ is

---

1. A *global BDD* is referred to a BDD represented in terms of the primary inputs.

not zero. Despite of this problem, the experimental results indicate that a very high percentage of equivalent pairs can be identified without using the primary inputs as the cutset. In order to reduce the probability of false negative, a heuristic was developed that dynamically selects an appropriate cutset for each signal pair under equivalence checking. Because the discrepancy function is constructed only based on a local cutset in this approach, it can handle much larger designs than when global BDDs are used.

Constructing BDDs using a dynamic support

In the following the terms *cutset* and *support* are used interchangeably. Fig. 3.4 illustrates the heuristic for selecting the cutset. Given a signal pair

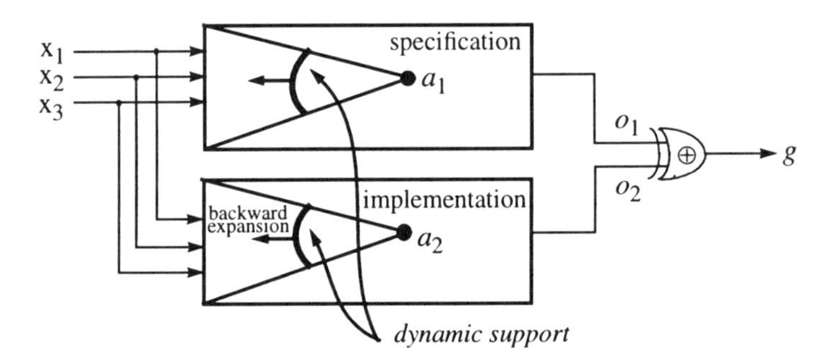

Fig. 3.4 The dynamic support that expands towards the primary inputs on demand for verifying the equivalence of (a_1, a_2).

(a_1, a_2), the heuristic first traverses the miter backward from signal a_1 and a_2 towards the primary inputs until it reaches *merge points or primary inputs*. The set of merge points and primary inputs where the traversal stops forms the first-level support λ_1. Using the first-level support as the cutset, the discrepancy function $Disc_{\lambda 1}(a_1, a_2)$ is constructed by computing $F_{\lambda 1}(a_1) \oplus F_{\lambda 1}(a_2)$. If $Disc_{\lambda 1}(a_1, a_2)$ is zero, then it can be concluded that the target pair is equivalent. Otherwise, we further traverse the miter backward from the first-level support toward primary inputs to find the second-level support. In principle, the support advances towards the primary inputs dynamically if a lower-level support is not sufficient to prove the equivalence of the target signal pair. Toward a similar objective, Bushnell et al. have used the idea of justification frontiers [22,29].

In the worst case, we need to advance the frontier of the support all the way to the primary inputs to decide whether a signal pair is indeed equivalent or not. This degenerates to the case of constructing global BDD. To avoid potential memory explosion, a user-specified limit on the levels of backward support expansion can be set. If a signal pair cannot be proven equivalent after reaching the limit, it can be treated as inequivalent pessimistically and the heuristic moves on to check the next candidate signal pair. The only exception to this strategy is when the target pair is a primary output pair. In that case, the heuristic switches to the ATPG technique to trade time for space and search for any possible distinguishing vector.

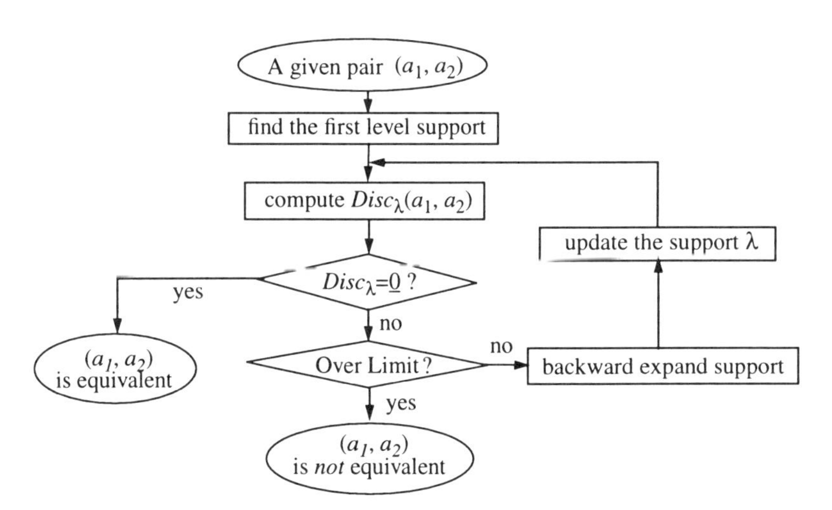

Fig. 3.5 Routine for checking the equivalence of a signal pair using local BDD with dynamic support.

The flow of checking equivalence for a signal pair is shown in Fig. 3.5. We expand the current support toward primary inputs for two logic levels each time when selecting the next support. Also, in an attempt to find a smaller cutset, we only select those merge points that have multiple fanouts as the supporting signals.

Example 3.2 Fig. 3.6 shows the snapshot of an example during the incremental verification process. Suppose signal pairs (a, a'), (b, b'), and (c, c') have been proven equivalent and merged together. The algorithm proceeds to check the equivalence of the primary output pair (o_1, o_2). Based on the cutset selection algorithm, $\lambda_1 = \{b', c'\}$ is the first level cutset as shown in Fig. 3.6(b). We compute $Disc_{\lambda 1}(a_1, a_2)$, which is $\{(b', c') \mid (1, 0)\}$. Because it is not empty, we further expand it backward to the second level cutset $\lambda_2 = \{x_1, a', x_3\}$. At this point, we find that $Disc_{\lambda 2}(a_1, a_2)$ is a tautology zero. Hence, we conclude that (o_1, o_2) is an equivalent pair. In this example, the distinguishing vector with respect to the first level cutset $(b', c') = (1, 0)$ is proven to be an impossible value combination when we advance the cutset one level towards the primary inputs, and thereby the false negative problem is resolved.

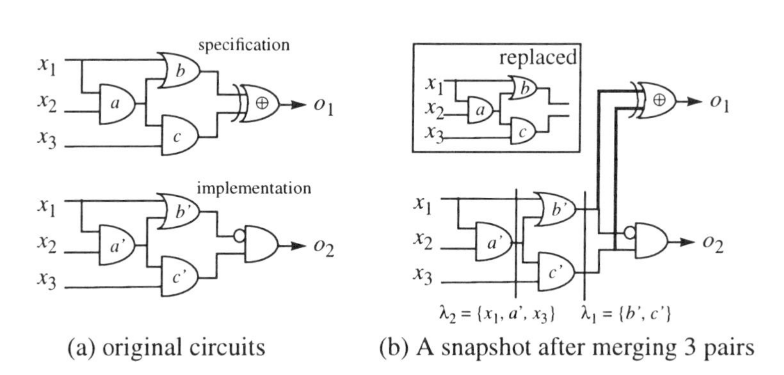

(a) original circuits (b) A snapshot after merging 3 pairs

Fig. 3.6 A snapshot as checking the equivalence of signal pair (o_1, o_2).

Experimental results

Table 3.1 shows the results of a program to verify the correctness of ISCAS-85 benchmark circuits minimized by *script.rugged* in SIS [121] and further mapped using library *mcnc.genlib*. These results are obtained by running the program on a Sun Sparc-20 workstation equipped with 128MB memory. It can be seen that every circuit is verified within one minute of CPU time (which includes the random simulation time), indicating this approach to be quite robust and applicable to circuits that have gone through intensive logic transformations. Experience shows that many equivalent pairs cannot be identified unless we backward expand the dynamic support for a number of levels as discussed earlier. In the worst case, this approach, if no limit were set, can degenerate to the global BDD-based approach. However, the memory explosion problem does not occur for the entire set of ISCAS-85 benchmark circuits because many internal equivalent pairs are identified. Columns 4 and 5 of Table 3.1 list the number of signal pairs with the same name, and the total number of identified equivalent pairs or complementary pairs. Here, the ratio of signal pairs of the same names to the total number of equivalent pairs is very small, indicating that the naming information was not useful in pairing up candidate signals.

Table 3.1 Results of verifying circuits minimized by *script.rugged* and then mapped into library cells defined in *mcnc.genlib* in SIS.

circuit	# nodes of C1	# nodes of C2	# pairs of same name	# equivalent signal-pairs (equivalent / complementary)	CPU-time (second)
C432	123	97	1	43 / 17	5
C499	162	184	0	101 / 53	1
C880	302	182	0	57 / 23	1
C1355	474	184	0	106 / 61	2
C1908	441	214	7	148 / 77	6
C2670	694	298	2	119 / 82	15
C3540	956	560	3	169 / 179	28
C5315	1454	661	3	271 / 153	10
C6288	2353	1419	1	849 / 334	10
C7552	2118	881	3	418 / 245	21

3.2 Learning-Based Algorithms

Around the same time when Brand proposed his incremental algorithm, Kunz and Stoffel also proposed a similar algorithm HANNIBAL [78,79] which uses *recursive learning* [77] to explore the internal similarity between the two circuits under verification.

3.2.1 Recursive Learning

Recursive learning can be viewed as a general method for solving the Boolean satisfiability problem and hence offers an alternative for logic verification[2]. In this section, we give a brief introduction to recursive learning. Consider the following example [12]. Suppose, we want to derive the mandatory assignments (or logical consequences) of assigning a '0' to signal a. Fig. 3.7(a) shows the mandatory assignments $\{b = 0, o_1 = 0\}$ produced by direct implication [49]. By using recursive learning, three more mandatory assignments, i.e., $\{c = 1, o_2 = 1, g = 0\}$, can be further derived as illustrated in Fig. 3.7(b). The procedure of recursive learning for deriving

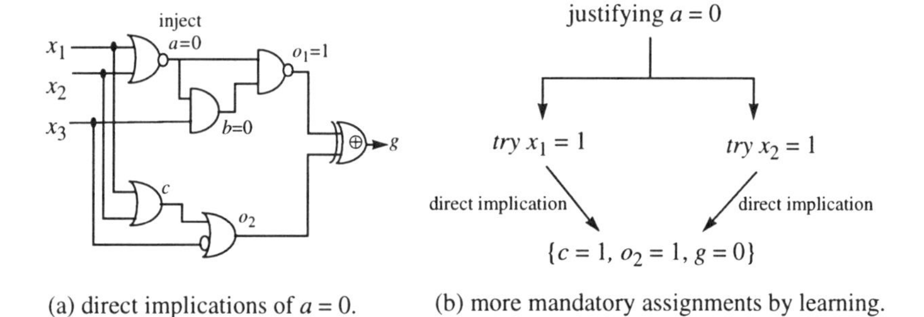

(a) direct implications of $a = 0$. (b) more mandatory assignments by learning.

Fig. 3.7 Example for recursive learning.

---

2. Checking the equivalence of (o_1, o_2) can be done by solving the Boolean equation:
$o_1(x_1, x_2, \ldots, x_n) \oplus o_2(x_1, x_2, \ldots, x_n) = 1$

these extra mandatory assignments is summarized as follows:

- Step 1: Inject '0' to signal a and perform direct implication.
- Step 2: Select an *unjustified gate*. A gate G is called *unjustified* "if there are unspecified input or output signals of G for which a combination of value assignments yielding a conflict at G exists". Unjustified gates describe the locations where *learning* is performed [77]. In this example, the logic gate at signal a is selected.
- Step 3: Enumerate possible combinations of value assignments for justifying the selected unjustified gate. In this example, in order to justify $a = 0$, there are two possible combinations, which are $\{x_1=1\}$ and $\{x_2=1\}$. Each of these possible combinations is called a *justification* for the selected unjustified gate.
- Step 4: Perform direct implication to collect the logical consequences for each justification: $\{x_1=1\}$ implies $\{c = 1, o_2 = 1, g = 0\}$, and $\{x_2=1\}$ implies $\{c = 1, o_2 = 1, g = 0\}$.
- Step 5: Take the *consensus* of every justification's logical consequences as the new mandatory assignments, which is $\{c = 1, o_2 = 1, g = 0\}$.
- Step 6: Update the mandatory assignments, which include $\{b = 0, o_1 = 0, c = 1, o_2 = 1, g = 0\}$.
- repeat Steps 2-6 recursively until no unjustified gate exists.

Since the complete recursive learning process could be very time-consuming, in many applications it is controlled by a limit on the *recursion depth*. Once the recursion depth exceeds this limit, the process terminates and returns a subset of mandatory assignments.

3.2.2 Verification Flow Using Recursive Learning

Fig. 3.8 shows the flow of the learning-based algorithm. Each iteration of this iterative process consists of two phases. The first phase identifies and stores the implications by sweeping the miter from primary inputs toward primary outputs using recursive learning. At each gate F, the algorithm assigns a *controlling value*, c, (the value that makes a gate unjustified) to the gate's output signal, f. Then a function called *make_all_implications(f, c, d)* is used to derive the implications of $f = c$ *by*

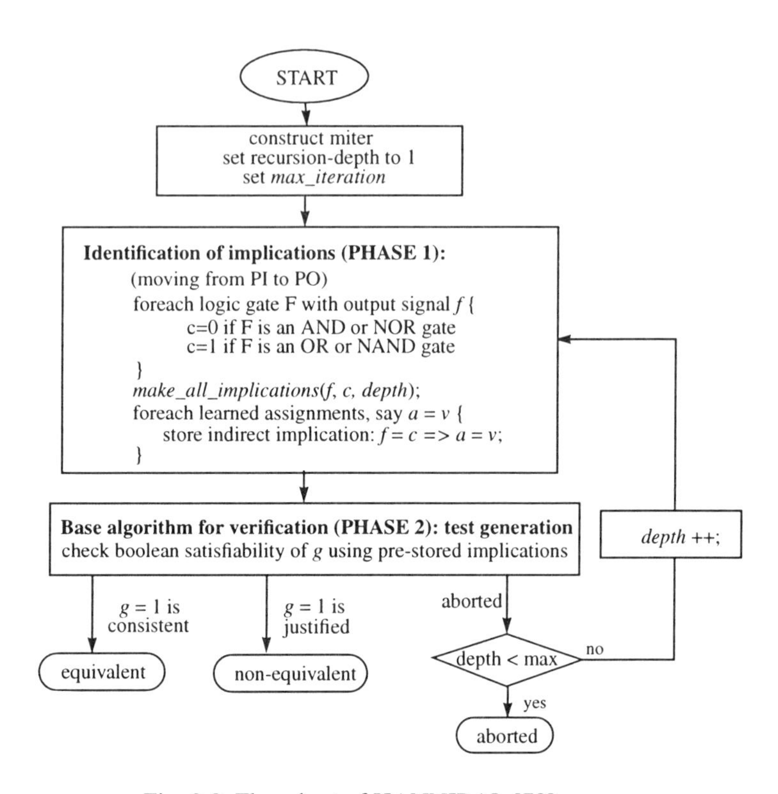

Fig. 3.8 Flowchart of HANNIBAL [78].

performing recursive learning with recursion depth, d. The recursion depth is initially assigned to 0, and then increased gradually, if needed, at each iteration. Every implication learned in this pre-processing phase is stored in the data structure of the target gate for use in the second phase. The second phase checks the satisfiability of the miter's output signal g. A test generator using *make_all_implications*() instead of a decision tree is applied to check satisfiability, so that the pre-stored implications in phase 1 can be easily utilized. If the satisfiability problem cannot be solved by the "*base*" algorithm with the existing implications, then phase 1 is called again with the depth of recursion augmented to a higher level value for deriving additional indirect implications. This process continues until the two CUVs are proven equivalent, inequivalent, or the limit of the maximum iterations set by the user is reached.

3.3 Transformation-Based Algorithm

This approach [110] performs logic transformations on the CUVs to enhance the structural similarity between the two CUVs in addition to just exploiting *existing* similarity. In this algorithm, the similarity is measured by the amount of indirect implications derived by recursive learning. The transformations designed to enhance the similarity are performed on signals in the region where the specification and the implementation are *dissimilar*. The specification as well as the implementation are actually augmented progressively during this process, instead of being simplified like the substitution-based methods. Although the miter becomes larger and larger, the two circuits under verification become more similar, and thus, the verification process becomes easier. The flow of this algorithm is shown in Fig. 3.9.

At each iteration of this process, the learning-based verification technique is first applied. As described in the previous subsection, that consists of two phases: learning phase and ATPG phase. If the ATPG phase fails to prove or disprove the circuit equivalence within pre-defined time budget, then the similarity enhancing transformations (SETs) are performed on those identified dissimilar regions before starting the next iteration with a higher recursion level. Two most important issues for this enhancement are: (1) how to identify the dissimilar regions, and (2) what transformations should be performed to increase the similarity.

3.3.1 Identifying Dissimilar Region

The dissimilar region is identified by a measure called *similarity index*. The similarity index of a signal a in C_1 (C_2) is the number of signals b_i in C_2 (C_1), such that a implies b_i or b_i implies a.

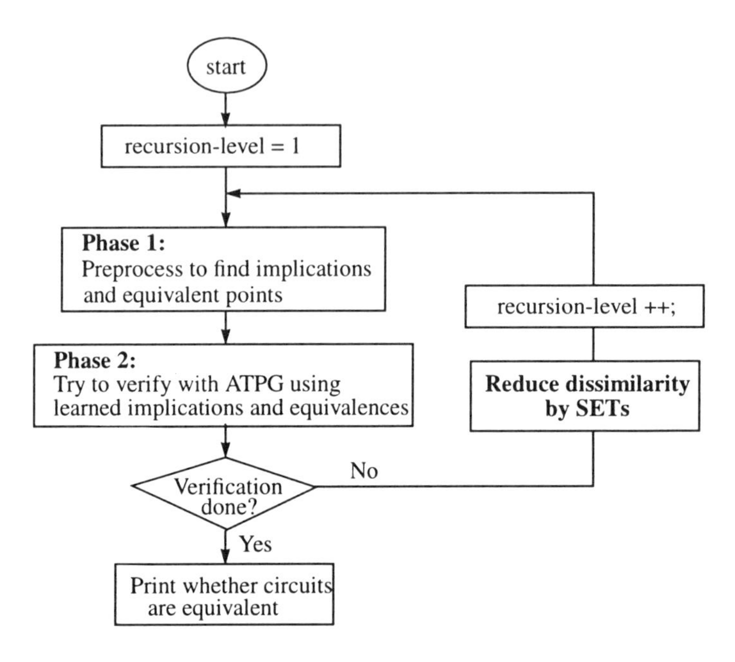

Fig. 3.9 Flowchart of transformation-based verification [110].

Example 3.3 Consider the two circuits in Fig. 3.10. Using level 1 recursive learning, there are three indirect implications associated with signal c:

$$(\boldsymbol{c} \text{ in } C_1) = 0 \text{ implies } (r \text{ in } C_2) = 0$$
$$(r \text{ in } C_2) = 0 \text{ implies } (\boldsymbol{c} \text{ in } C_1) = 0$$
$$(\boldsymbol{c} \text{ in } C_1) = 0 \text{ implies } (o_2 \text{ in } C_2) = 0$$

Thus, the similarity index of signal c is 3. Given a set of implications among the signals in the miter, the dissimilar region is defined as the set of signals with a zero similarity index. The set of similarity indices of all signals in the miter is called the *similarity profile*. In the example shown in Fig. 3.10, internal signals d and h in C_1, and r and v in C_2 constitute the dissimilar region. Since dissimilar region is identified using a given level of recursion in the recursive learning procedure, it is possible that a signal in the dissimilar region using some level of recursion may be found in the similar region using a higher level of recursion.

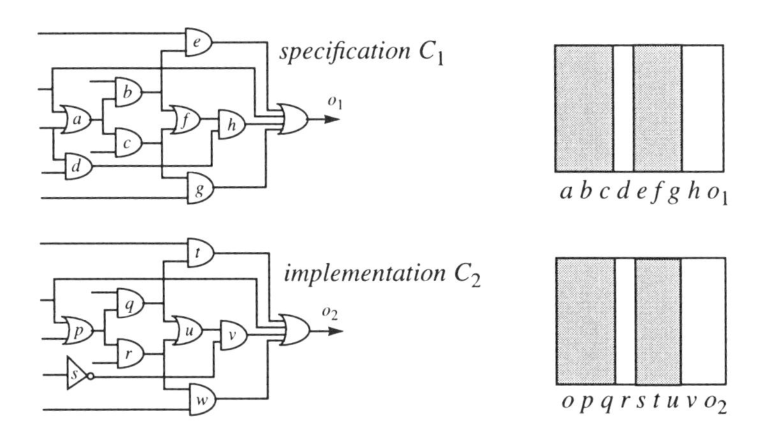

Fig. 3.10 Similarity profile.

3.3.2 Similarity Enhancing Transformation (SET)

Once the dissimilar region that represents the bottleneck of the learning-based verification phase is identified, the SET transformations designed to induce more similarity are performed on that region. As illustrated in Fig. 3.11, suppose signals f and h are in the dissimilar region of C_1 and C_2, respectively, the function of signal h can be replaced by a new function $h^{new} = \lambda(f, h)$ without changing the functions of the primary outputs based on the following theorem.

Theorem 3.1 Consider Fig. 3.11. Replacing h by a new function $h^{new} = \lambda(f, h)$ is a safe transformation for each of the following four cases:

$$(1)\ \lambda(f, g) = f + g \quad if \quad \left(f \cdot \bar{h} \cdot \frac{d(o_2)}{dh}\right) = 0$$

$$(2)\ \lambda(f, g) = f \cdot g \quad if \quad \left(\bar{f} \cdot h \cdot \frac{d(o_2)}{dh}\right) = 0$$

$$(3)\ \lambda(f, g) = \bar{f} + g \quad if \quad \left(\bar{f} \cdot \bar{h} \cdot \frac{d(o_2)}{dh}\right) = 0$$

$$(4)\ \lambda(f, g) = \bar{f} \cdot g \quad if \quad \left(f \cdot h \cdot \frac{d(o_2)}{dh}\right) = 0$$

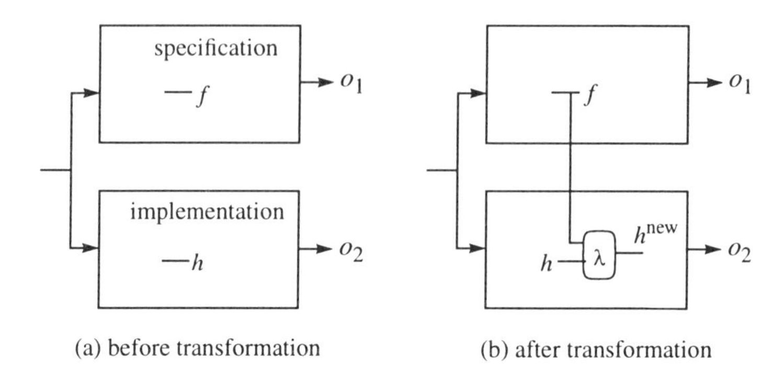

Fig. 3.11 Similarity Enhancing Transformation in a miter.

Explanation: Consider the first case when λ is an OR gate. The mite before the transformation is equivalent to the miter after transformation with the f stuck-at-0 fault. Therefore, the transformation is functionality-preserving *if and only if* the *f stuck-at-0 fault in the miter after transformation is redundant*. In other words, the transformation is safe if and only if there exists no input vector that can activate the fault (i.e., set f to '1') and propagate the fault effect through the added OR-gate to the primary output o_2 (i.e., set h to '0' and satisfy the Boolean difference of o_2 with respect to h). Taking the conjunction of these conditions, we arrive at $\left(f \cdot \bar{h} \cdot \frac{d(o_2)}{dh}\right) = 0$. The validity of these transformations has also been discussed in the context of logic minimization [28]. From another viewpoint, $\left(f \cdot \bar{h} \cdot \frac{d(o_2)}{dh}\right) = 0$ means that for all input vectors that make the effect of change at h observable at the primary output o_2 (i.e., $\frac{d(o_2)}{dh} = 1$), $f = 1$ must imply $h = 1$ (i.e., $(f \cdot \bar{h}) = 0$). In other words, a SET transformation can be viewed as a transformation utilizing the implications under the observability don't care conditions, $\frac{d(o_2)}{dh} = 0$. The reasonings for other cases are similar.

Given two signals f and h, checking if there exists a valid SET transformation between f and h (satisfying one of the four conditions in Theorem 3.1) involves complete ATPG. To reduce the complexity, techniques based on mandatory assignments can be applied. For example, if $h = 1$ is a mandatory assignment for detecting f stuck-at-0, then $h = 0$ and $\left(f \cdot \frac{d(o_2)}{dh}\right) = 1$ cannot be satisfied at the same time, and thus, an OR-gate SET transformation (case 1 of Theorem 3.1) is valid at signal h. The detail of this technique can be found in the literature [28,79].

Example 3.4 Consider a SET transformation at signal c and s as shown in Fig. 3.12. An OR-gate has been added at signal s. The new signal is denoted as z. Compared to the miter before the transformation shown in Fig. 3.10, several new additional implications have been induced. In fact, this SET alone induces the final implications between the primary outputs of C_1 and C_2 (i.e., $o_1 = 0$ implies $o_2 = 0$, and $o_1 = 1$ implies $o_2 = 1$).

3.4 Summary

We have discussed three types of incremental approaches:

- Substitution-based approach [13,62,97]: sweeping the miter from primary inputs (PI's) towards primary outputs (PO's) to identify equivalent signal pairs or permissible signal pairs. Once an equivalent or a permissible pair is found, the two signals are merged (i.e., one signal is replaced by another) to reduce the size of the miter and the complexity of the subsequent equivalence checking for other signal pairs and primary output pairs.
- Learning-based [71,78,112]: exploring and recording the correlations among signals from PI's towards PO's. The correlations are recorded in the circuit's data structure and then utilized subsequently to prove the equivalence of each corresponding primary output pair of the two CUV's.
- Transformation-based [110]: incorporating logic transformation techniques to transform the miter in such a way that more similarity

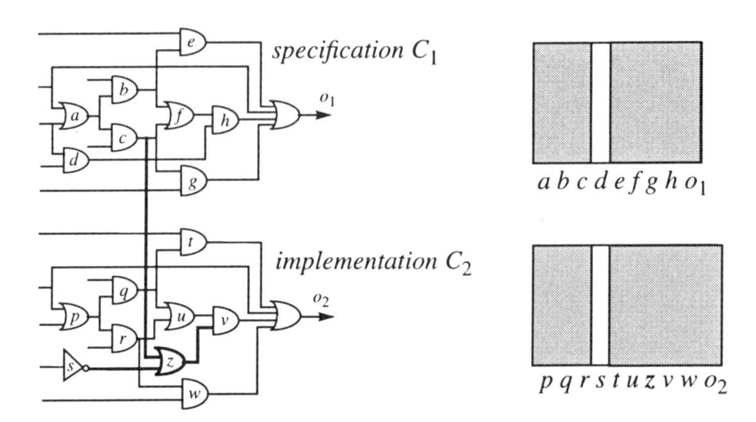

Fig. 3.12 Effect of similarity profile after a SET transformation.

between CUV's can be created or explored with relative ease in the subsequent process.

These approaches differ from one another in the types of similarity being utilized. For example, substitution-based approaches take advantage of the permissible pairs between CUV's, but do not use the one-way implications (i.e., a=0 implies b=0, but not vice versa). On the contrary, learning-based approaches take advantage of the one-way implications, but do not consider the permissible pairs.

Incremental approaches also vary from one another in the techniques applied to explore the similarity. Automatic Test Pattern Generation (ATPG) and local BDD are two most popular techniques. ATPG is often used as a Boolean reasoning engine through branch-and-bound search to decide if a given Boolean objective is satisfiable. When used as a means to prove equivalence of a signal pair (could be an internal or a primary output pair), the entire input space is exhausted to make sure that no input vector can differentiate the target signal pair. Information of the signal correlations can be used to trim down the search space dramatically. On the other hand, local BDD is a technique that constructs the BDD representation with internal signals as the supporting variables, not necessarily the primary inputs.

As has been reported [62,97], the local BDD-based techniques are more efficient than a pure ATPG-based method for a wide range of test cases. However, the former is also more memory intensive. In the worst case, it could degenerate to the traditional symbolic approach, and thus, is relatively more vulnerable to memory explosion. On the other hand, a ATPG-based method, although may be more time-consuming for checking equivalence, is particularly effective in finding a counter-example when the two CUV's are inequivalent.

To conclude this chapter, we summarize the incremental algorithms in Table 3.2 based on the criteria of:

(1) the types of similarities explored,

(2) the underlying techniques used to explore the similarity, and

(3) the utilization of the identified similarity.

Table 3.2 A comparison of incremental algorithms for checking combinational equivalence.

Algorithms	Similarity Explored	Exploring Techniques	Utilization
[Brand-93]	permissible pairs	ATPG	substitution
[Kunz-93]	implications	recursive learning	learning
[Jain-95]	implications	functional learning	learning
[Reddy-et-al-95]	implications	recur. learning + BDD	hybrid
[Huang-et-al-96]	equivalent pairs	local BDD	substitution
[Matunaga-96]	equivalent pairs	local BDD	substitution
[Pradhan-et-al-96]	implic. with ODC	recursive learning	transformation
[Kuehlmann-et-al-97]	equivalent pairs	local BDD	substitution

There are four types of similarities being explored: permissible pairs, equivalent pairs only, implications, and enhanced implications with observability don't cares (ODC). In terms of the techniques exploring the similarity, there are: ATPG, recursive learning, functional learning, and local BDD. Finally, in terms of how the identified similarity is utilized, these approaches can be classified into three types: substitution-based, learning-based, and transformation-based.

Chapter 4

Incremental Verification for Sequential Circuits

In this chapter we address the problem of verifying the equivalence of two sequential circuits. In an attempt to handle larger circuits, we modify the test pattern generation technique for verification. The suggested approach utilizes the efficient backward justification technique popularly used in most sequential ATPG programs. We present several techniques to enhance the efficiency of this approach: (1) identifying equivalent flip-flop pairs using an induction-based algorithm, and (2) generalizing the idea of exploring the structural similarity between circuits to perform verification in stages. This ATPG-based framework is suitable for verifying circuits either with or without a reset state. Experimental results on verifying the correctness of circuits after sequential redundancy removal with up to several hundred flip-flops are presented.

This chapter is organized as follows. In Section 4.1, we discuss and compare several definitions of sequential equivalence. In Section 4.2, we describe a general ATPG-based framework for verifying circuit equivalence according to different definitions. In Section 4.3, we introduce several speed-up techniques. In Section 4.4, we present the experimental results. In Section 4.5, we give a summary.

4.1 Definition of Equivalence

A synchronous sequential circuit can be described in terms of its corresponding deterministic finite state machine (DFSM). Assume two circuits under consideration are $M_1 = (I, O, S_1, \delta_1, \lambda_1)$ and $M_2 = (I, O, S_2, \delta_2, \lambda_2)$, respectively, where I and O are the input and output alphabets. S is a set of states, δ is the transition function from $S \times I$ to S, and λ is the output function from $S \times I$ to O. A *synchronizing sequence* is an input sequence that can bring the machine to a unique state from any state, while an *initializing sequence* is a synchronizing sequence that can be verified using 3-valued logic simulation.

4.1.1 Equivalence of Circuits With A Reset State

Definition 4.1 (*equivalent state pair*) Let s_1 be a state in M_1, and s_2 be a state in M_2. State s_1 is equivalent to state s_2, denoted as $s_1 \sim s_2$, if there exists no input sequence that can distinguish this state pair, i.e., for any input sequence $\pi \in I$, $\lambda_1(s_1, \pi) = \lambda_2(s_2, \pi)$, where λ_1 and λ_2 are the output functions for M_1 and M_2, respectively.

Definition 4.2 (*reset equivalence*) Suppose M_1 and M_2 have an external reset line for every flip-flop and the reset states are denoted as s_1 and s_2, respectively. M_1 and M_2 are called reset equivalent if and only if s_1 and s_2 are equivalent states.

4.1.2 Equivalence of Circuits Without A Reset State

For circuits without an external reset state, several notions of sequential equivalence have been proposed [32,104,105,107]. These definitions differ from one another in the assumption of how a circuit normally operates. In the following, we discuss four definitions of equivalence for circuits without a known reset state: (1) sequential hardware equivalence, (2) safe replaceability, (3) three-valued safe replaceability, and (4) three-valued equivalence.

Sequential Hardware Equivalence

For some circuits, a synchronizing sequence is used to drive the circuits to a desired state before operation. Thus, the notion of *sequential hardware equivalence* (SHE) is useful for such a circuit [104]. This definition primarily concerns the post-synchronization (or steady-state) behavior of a design. If two designs have the same behavior after synchronization, then they are considered to be equivalent. To interpret this notion graphically, consider the steady-state behavior of a synchronizable design as determined by one *Terminal Strongly Connected Component* (TSCC) in the state transition graph. A TSCC is a strongly connected component (SCC) that does not have any outgoing edge. It has been proven [104] that a transformation preserving the behavior of its TSCC does not change its input/ output behavior after synchronization (assuming that the transformed circuit can still be synchronized). Here, the equivalence of two TSCCs is based on the conventional definition of graph equivalence defined as follows.

Definition 4.3 (*graph equivalence*) Suppose T_1 and T_2 are two state transition graphs. These graphs are equivalent if and only if every state in T_1 has an equivalent state in T_2, and every state in T_2 has an equivalent state in T_1 as well.

Fig. 4.1 shows a transformation that preserves the TSCC. Under the definition of sequential hardware equivalence, this is a valid transformation. This notion of equivalence will also be referred to as *post-synchronization equivalence* in the sequel. In what follows, we define alignable state pair before introducing the formal definition of SHE.

Definition 4.4 (*alignable pair* [104]) (s_1, s_2) is an alignable (state) pair if there exists an input sequence π that can bring the two circuits into an equivalent state pair from (s_1, s_2), i.e., $\delta_1(s_1, \pi) \sim \delta_2(s_2, \pi)$, where δ_1 and δ_2 are the transition functions. The sequence π is called an aligning sequence for the state pair (s_1, s_2).

Definition 4.5 (*sequential hardware equivalence* [104]) M_1 and M_2 are equivalent if and only if there exists an input sequence π that can align *any*

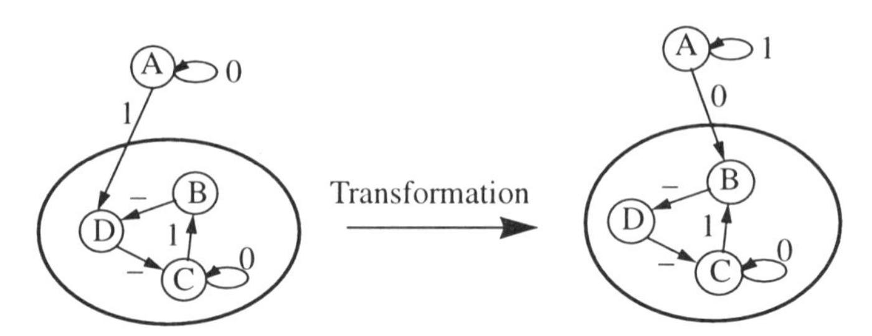

Fig. 4.1 A transformation that preserves Terminal Strongly Connected Component.

state pair in $S_1 \times S_2$ (π is called a universal aligning sequence).

A universal aligning sequence can bring two circuits into an equivalent state pair regardless of their initial states. The above definition states that the two circuits have the same post-synchronization behavior as long as there exists a universal aligning sequence. It can be shown that if the two circuits are sequential hardware equivalent, then *any* input sequence that can *synchronize* both circuits is a universal *aligning* sequence. Therefore, if a synchronization sequence is given for both circuits, checking the sequential hardware equivalence can be done by checking the reset equivalence using the following property.

Property 4.1 Let π_1 and π_2 be synchronizing sequences for M_1 and M_2, respectively. Note that the input sequence $\pi_1\pi_2$ (i.e., the concatenation of π_1 and π_2) is also a synchronizing sequence for both circuits. Suppose $\pi_1\pi_2$ brings the two circuits into states s_1 and s_2, respectively. Then M_1 and M_2 are sequential hardware equivalent if and only if (s_1, s_2) is an equivalent state pair.

Explanation: If (s_1, s_2) is an equivalent state pair, then by definition, two circuits are sequential hardware equivalent. The proof of the other way is less obvious. Suppose (s_1, s_2) is *not* an equivalent state pair. Then there exists an input sequence $\pi_1\pi_2$ that brings *any* state pair to an inequivalent state pair (s_1, s_2). In other words, input sequence $\pi_1\pi_2$, referred to as a *universal distinguishing sequence* in the sequel, can differentiate the two circuits regardless of their initial states. It follows that there exists no equivalent state pair and, thus, the two circuits are not sequential hardware equivalent.

Safe Replaceability

The notion of SHE may not be suitable for a circuit embedded in a larger design because it ignores the output responses before the circuit is initialized. Even if a transformation does not change the steady state behavior of the circuit and thus is valid under SHE, it may destroy the synchronizing sequence associated with the entire design because the transient behavior (i.e., the behavior before synchronization) of the embedded circuit under consideration could be changed. To overcome this limitation, another notion called the *safe replaceability* is proposed in [105]. The safe replaceability requires that the I/O behavior of the transformed circuit is contained in the I/O behavior of the original circuit. This ensures that the surrounding environment of the target circuit cannot detect any difference when the original circuit is replaced by the transformed circuit.

Definition 4.6 (*safe replaceability* [105]) M_2 is a safe replacement for M_1 if given any state s_2 in M_2 and any finite input sequence π, there exists some state s_1 in M_1 such that their output responses to π are identical, i.e., $\lambda_1(s_1, \pi) = \lambda_2(s_2, \pi)$.

It has been proven [105] that if M_2 is a safe replacement for M_1, then the Terminal Strongly Connected Components (TSCC) of M_1 and M_2 are isomorphic. That is, the safe replaceability is a more stringent definition than the sequential hardware equivalence because the former preserves the behavior of the TSCC, which defines the post-synchronization behavior. However, checking safe replaceability is not an easy task because it involves the enumeration of each state in M_2.

Three-Valued Safe Replaceability

Three-valued safe replaceability is another definition of sequential equivalence for circuits without a reset state. This definition assumes that the initial value of each flip-flop is unknown and represented as 'u'. For an input sequence applied to the circuit, the three-valued logic simulation is used to derive the final response of each internal signal and primary output. If the response of a primary output bit is '0' or '1', then the output response is regarded as a *care* output response and should be preserved in the transformed circuit to satisfy this definition. On the other hand, if the response is an 'u' (it means that the response could be '0' or '1' depending on the power-up values of flip-flops), then it is regarded as a *don't care response* and could be either maintained or changed in the transformed circuit. This definition has been used to describe the behavior of an incompletely specified FSM [98]. We will show that this definition is more stringent than sequential hardware equivalence if the original circuit is 3-valued initializable. A circuit is called 3-valued initializable if it can be initialized to a unique state by the 3-valued logic simulation using a synchronizing sequence [31,118].

The three-valued safe replaceability is useful if the specification is verified by the 3-valued logic simulation. Three-valued logic simulation is known to be conservative in the sense that a simulated signal may be represented as a 'u' when it is actually a '0' or a '1' regardless of the power-up state [2]. However, due to the lack of other fast methods to compute a circuit's output response, most designers still rely on 3-valued logic simulation to verify their design. In other words, a design is considered as correct only if the result of the 3-valued logic simulation agrees with the desired behavior.

In the context of verifying 3-valued safe replaceability, we assume a signal in the circuit can take on a logic value among $\{0, 1, u\}$, where u represents the unknown logic value. We define that signal v_1 covers signal v_2 if (v_1, v_2) is one of the following 5 value combinations: (u, u), $(u, 0)$, $(u, 1)$, $(0, 0)$, $(1, 1)$. In other words, (v_1, v_2) is not one of the following 4 combinations: $(0, 1)$, $(1, 0)$, $(0, u)$ and $(1, u)$. Similarly, a vector V_1 covers vector V_2 if every bit of V_1 covers the corresponding bit of V_2. Each signal in the combinational portion of a circuit is a function of primary inputs PI = $\{i_1, i_2,..., i_n\}$ and present state lines PS = $\{y_1, y_2,..., y_m\}$. In a 3-valued state (or a state cube), each state line could be 0, 1 or u. The unknown state, denoted as x, is a 3-valued state where every bit is u, e.g., $x = (uuu)$. Suppose T is an input sequence, the primary output response (at the last cycle of the sequence) of the circuit M_1 after applying T from a 3-valued state s is denoted as $o_1(s, T)$. Similarly, $o_1(x, T)$ denotes the output response of o_1 from the unknown state x.

Definition 4.7 (*three-valued safe replaceability*) Circuit M_2 is a 3-valued safe replacement for circuit M_1 (denoted as $M_1 \subset M_2$) if and only if $o_1(x, T)$ covers $o_2(x, T)$ for any input sequence T.

The above definition applies to a circuit operated from an unknown initial state. It can be shown that if M_2 is a 3-valued safe replacement for M_1, then all initializing sequences of M_1 can also initialize M_2.

Three-Valued Equivalence

Definition 4.8 (*three-valued equivalence*) M_1 is 3-valued equivalent to M_2 if and only if for all input sequences T, the output combination $(o_1(x, T), o_2(x, T))$ belongs to $\{(0, 0), (1, 1), (u, u)\}$.

This definition states that M_1 and M_2 are 3-valued equivalent if and only if their output responses are identical for any input sequence assuming that both circuits start from the unknown state, that is, the other 6 output combinations, $\{(0, 1), (1, 0), (u, 0), (u, 1), (0, u), (1, u)\}$, are not allowed for any primary output pair with respect to any input sequence T. Obviously, if M_1 is 3-valued equivalent to M_2, then M_2 is a 3-valued safe replacement of M_1, but not vice-versa.

4.1.3 Comparison of Definitions

If the specification M_1 is 3-valued initializable, it can be proven that M_1 is sequential hardware equivalent to any of its 3-valued safe replacement. In other words, 3-valued safe replaceability is a more stringent definition than SHE for most practical designs where a 3-valued initializing sequence exists, even though the 3-valued safe replaceability is defined over the conservative 3-valued logic simulation. Before proving this property, we first discuss two other relevant properties of 3-valued safe replaceability. First, it is *3-valued compositional*. Second, it *preserves every 3-valued initialization sequence* of the specification circuit in any of its 3-valued safe replacement. These three properties jointly make the 3-valued safe replaceability a viable definition for an embedded circuit as will be discussed later.

Property 4.2 (*3-valued safe replaceability is compositional*) Consider a network consisting of a number of sub-networks. If one or several subnetworks are replaced by their 3-valued safe replacements, then the resulting network is a 3-valued safe replacement of the original network.

Explanation: We illustrate this property by an example in Fig. 4.2. The network consists of two sub-networks, denoted as Ω_1 and Ω_2, that are later replaced by their 3-valued safe replacements Ω_1' and Ω_2', respectively. As shown in Fig. 4.2, the primary inputs and outputs of the network are denoted as $\{x_1, x_2, x_3\}$ and $\{z_1, z_2, z_3\}$. The interconnections between the two sub-networks are denoted as $\{y_1, y_2, y_3\}$.

We define a 3-valued vector V_1 as a *refinement* of another vector V_2 if V_1 is covered by V_2 bit-by-bit. For example, vector $(y_1, y_2, y_3) = (0, 0, 0)$ is a refinement of $(y_1, y_2, y_3) = (u, u, 0)$. In Fig. 4.2, we show that the 3-valued safe replaceability is compositional by asserting that, for any input sequence simultaneously applied to both M_1 and M_2, the output responses of the transformed design M_2 at $\{z_1, z_2, z_3\}$ is a refinement of the output responses of the original design M_1 at $\{z_1, z_2, z_3\}$. By definition, the responses of M_2 at the interconnections $\{y_1, y_2, y_3\}$ is a refinement of the responses of M_1 at $\{y_1, y_2, y_3\}$ because Ω_1' is a 3-valued safe replacement of Ω_1. With this condition, it can be further proven that the output response of M_2 is a refinement of M_1 at $\{z_1, z_2, z_3\}$ based on two arguments. First, Ω_2' is a 3-valued safe replacement of Ω_2. Second, the traditional 3-valued logic simulation is a response computation method that satisfies the

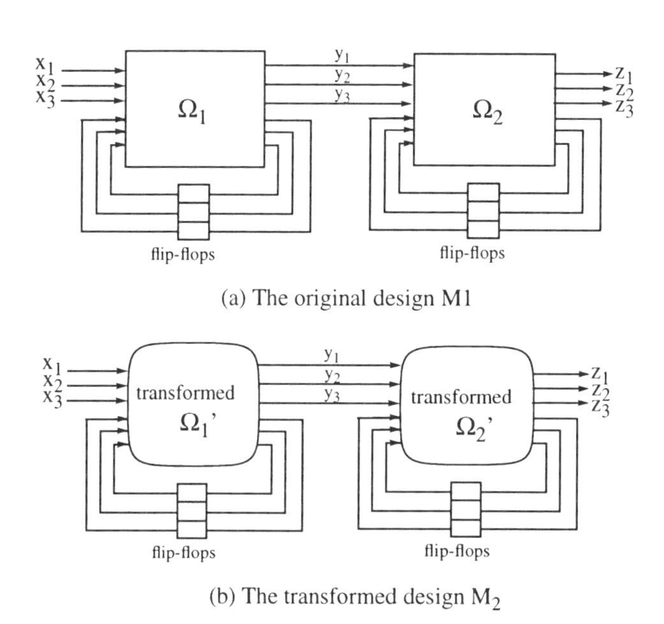

Fig. 4.2 An example showing that 3-valued safe replaceability is compositional.

criterion called *monotonicity* proposed by Brand et al. [15]. Monotonicity requires that a response computation method (or simulation method) produces *refined output responses when the input vector to a design is refined*. That is, for an output signal f, $f(V_2)$ is covered by $f(V_1)$ if V_2 is covered by V_1, where $f(V_2)$ represents the output response of f to the input vector V_2 according to a particular response computation method.

At each clock cycle, the input vector to Ω_2' is a refinement of the input to Ω_2 at $\{y_1, y_2, y_3\}$. By monotonicity and the condition that Ω_2' is a 3-valued safe replacement of Ω_2, it follows that the output response of M_2 is a refinement of M_1 at $\{z_1, z_2, z_3\}$ and, thus, 3-valued safe replaceability is compositional. It can be shown that this argument is also true for a complex design consisting of a number of sub-networks that are interconnected arbitrarily. It should be pointed out that SHE is not a compositional.

Property 4.3 (*3-valued safe replaceability preserves any initialization sequence*) Suppose M_2 is a 3-valued safe replacement for M_1, and π initializes M_1. Then π also initializes M_2.

Explanation: Because the output responses of M_1 after applying π are all binary and deterministic ('0' or '1'), the output responses of M_2 after π should therefore be binary and deterministic, and thus we conclude that π also initializes[1] M_2.

The above properties suggest the following advantage of using the 3-valued safe replaceability. A large initializable network can be optimized by replacing one sub-network at a time without looking at the sub-network's surrounding environment. The revised design can still be initialized by any initialization sequence[2] for the original network. Furthermore, the *post-synchronization* behavior is preserved.

Lemma 4.1 If M_1 is 3-valued initializable and M_2 is a 3-valued safe replacement for M_1, then M_2 is sequential hardware equivalent to M_2.

Proof: Let π initialize M_1, and M_2 be a 3-valued safe replacement for M_1. By Property 4.3, it follows that π initializes M_1. If M_1 and M_2 are *not* sequential hardware equivalent, then there exists a universal distinguishing sequence π_D that can differentiate any state pair (Property 4.1). Therefore, the output combination (o_1, o_2) in response to π_D from the unknown state using the 3-valued logic simulation is either (0, 1) or (1, 0). Thus, it violates the given condition that M_2 is a 3-valued safe replacement for M_1. Hence, by contraposition, we conclude the lemma. (Q.E.D.)

Another comparison of SHE and the 3-value safe replaceability can be made in the context of stuck-at fault removal process. Methods are proposed in [7,69,107] to remove untestable yet possibly irredundant faults in a sequential circuit without an external reset state for achieving a higher

1. An input sequence is called a *weakly initialization sequence* [107] if it brings the machine into a set of equivalent states, instead of a unique state. In this chapter, we make no difference between initialization and weak initialization because the circuit will start producing all binary outputs after the application of either one of them.
2. According to Property 4.1, any initialization sequence will bring the original and the revised network into equivalent states.

fault coverage. A fault is called *partially testable* if there exists an input sequence T such that the output combination of the fault-free and faulty circuits for this sequence is (0, u) or (1, u) and there exists no input sequence producing output combination (0, 1) or (1, 0). Removing such faults will violate the condition of 3-valued safe replaceability (because the partial test sequence serves as a counter-example). However, it is proven that removing this kind of fault preserves the post-synchronization behavior as long as the transformed circuit is still synchronizable. [69,107]. The faulty circuit on the right-hand side of Fig. 4.3 [107] is obtained by injecting a stuck-at-1 fault into the first literal (which is x) of the second product term of Y_2. This fault is partially testable because the output

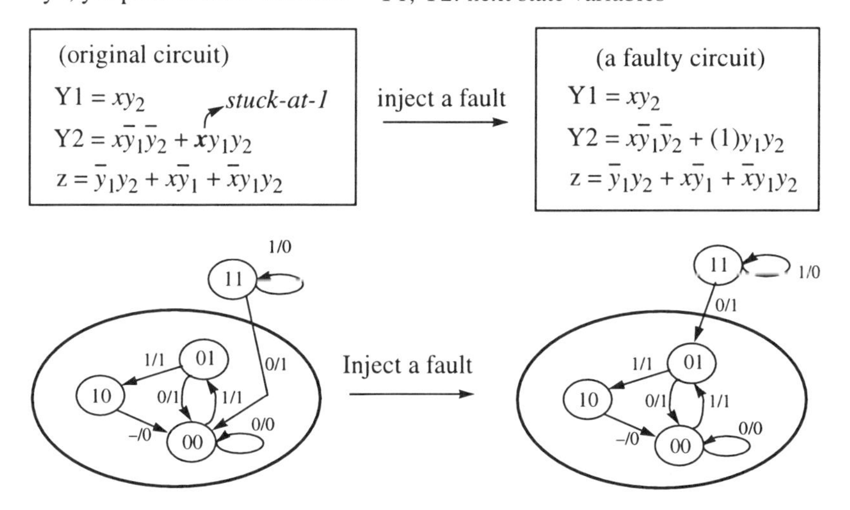

Fig. 4.3 An example to show that some transformation satisfying sequential hardware equivalence may violate 3-valued safe replaceability.

responses of the fault-free and faulty circuits for input sequence ($x = 0$, $x = 0$) are '0' and 'u' derived by the 3-valued logic simulation. Therefore,

($x = 0$, $x = 0$) is a 3-valued distinguishing sequence, and thus, the faulty circuit is not a 3-valued safe replacement of the fault-free circuit. On the other hand, ($x = 0$, $x = 0$) is a 3-valued initialization sequence that brings both circuits into an equivalence state pair $(Y_1 Y_2) = (00)$ in both the fault-free circuit and faulty circuit. Hence, they are sequential hardware equivalent.

Based on Lemma 4.1 and the above example, we thus conclude that the 3-valued safe replaceability is a more stringent definition than sequential hardware equivalence if the original circuit is 3-value initializable. In light of this, it is our conjecture that even though removing a partially testable fault that does not destroy all synchronizing sequences is a safe transformation with respect to the definition of sequential hardware equivalence, it may not be suitable for an embedded design or for a design producing *care outputs* before being completely initialized.

4.2 Methodology

In this section, a method to verify sequential equivalence of two circuits is discussed. This approach can handle circuits with or without a reset state. For circuits without a reset state, we verify them with respect to the definition of 3-valued safe replaceability. We construct a miter that connects the primary inputs of two circuits together, and connects each primary output pair to an XOR gate as shown in Chapter 1. Without loss of generality, we assume both circuits have only one primary output for the rest of this chapter.

Efficient ATPG techniques are applied to search for a test for the stuck-at-0 fault at the miter's output. For circuits with a reset state, untestability of this fault implies that the two circuits under verification are reset equivalent. But for circuits without a reset state, additional conditions need to be checked for *3-valued safe replaceability*. The underlying ATPG techniques rely on the time-frame-expansion model and the reverse time processing techniques [35,95], popularly used in many commercial ATPG tools.

4.2.1 Checking Three-Valued Safe Replaceability

Definition 4.9 (*untestability*) g stuck-at-0 fault is called *untestable* if there exists no input sequence T such that $g(x, T) = 1$, or equivalently, $(o_1(x, T), o_2(x, T))$ is either (0, 1) or (1, 0), where x represents the unknown state.

The untestability of g stuck-at-0 is simply a necessary condition for the 3-valued safe replaceability. Even if g stuck-at-0 fault is untestable, there may still exist an input sequence T such that $(o_1(x, T), o_2(x, T))$ is either $(0, u)$ or $(1, u)$ and violates the 3-valued safe replaceability. The following lemma gives the necessary and sufficient condition for the 3-valued safe replaceability.

Lemma 4.2 M_2 is a 3-valued safe replacement for M_1 if and only if there exists no input sequence T and state s_2 in M_2 such that $(o_1(x, T), o_2(s_2, T))$ belongs to $\{(0, 1), (1, 0)\}$.

Proof: See Section 4.6.

In the above lemma, if s_2 is the unknown state x, then the input sequence T is a test for g stuck-at-0 fault of the miter. Otherwise, T is called a partial test. In the following discussion, T is also called a distinguishing sequence for M_1 and M_2. The above necessary and sufficient condition for the 3-valued safe replaceability can be checked by a sequential ATPG program with very minor modification. A sequential ATPG program which employs the time-frame-expansion model, consists of three phases: (1) fault injection, (2) fault-effect propagation, and (3) backward justification. In this application, the fault is at the output of the miter and, thus, no fault effect propagation is required. Only the backward justification is needed after a fault is injected (i.e., asserting signal g to 0). The backward justification process may take more than one time frame to complete. At each time frame, the set of binary values required at the present-state lines is called a "state requirement", which needs to be further justified until the above necessary and sufficient condition is either satisfied or proven unjustifiable. A state requirement of the miter can be regarded as a 3-valued state consisting of two parts: the state requirement for M_1, and the state requirement for M_2. For example, $(0uu \mid 1uu)$ is a state requirement that requires a '0' for M_1's first state variable, and a '1' for M_2's first state variable.

According to Lemma 4.2, the backward justification process stops whenever one of the following conditions is satisfied:

(1) *Unjustifiable condition*: all state requirements generated during the search of a distinguishing sequence are proven unjustifiable. Then M_2 is a 3-valued safe replacement for M_1.

(2) *Justified condition*: a state requirement that does not have a requirement on M_1 is reached. For example, $(uuu \mid 1uu)$ is a state requirement that has no requirement on M_1. It implies that a distinguishing sequence is found and M_2 is not a 3-valued safe replacement for M_1.

Some ATPG algorithms such as those based on PODEM [54] may over-specify the value requirements at certain signals to speed up the search. It should be pointed out that the underlying ATPG algorithm used for verification must not over-specify the value requirements in the backward justification for checking the 3-valued safe replaceability. Otherwise some possible distinguishing sequences may be overlooked. There are techniques [34] that resolve the over-specification problem. Fig. 4.4 shows an example in which a distinguishing sequence, $(t_3\ t_2\ t_1)$, is found after 3 time frames are expanded for backward justification. The state requirement of the left-most time frame does not have any value requirement for M_1's state variables, and thus the search stops.

4.2.2 Checking Reset Equivalence

To check reset equivalence, the search for a sequence that can distinguish two circuits from the initial state pair of M_1 and M_2 is performed. Assume that the initial states of M_1 and M_2 are s_1 and s_2. An input sequence T is a distinguishing sequence if $(o_1(s_1, T), o_2(s_2, T))$ belongs to $\{(0, 1), (1, 0)\}$. This can be checked by modifying the stopping criteria of the backward justification process. In this case, the justified condition

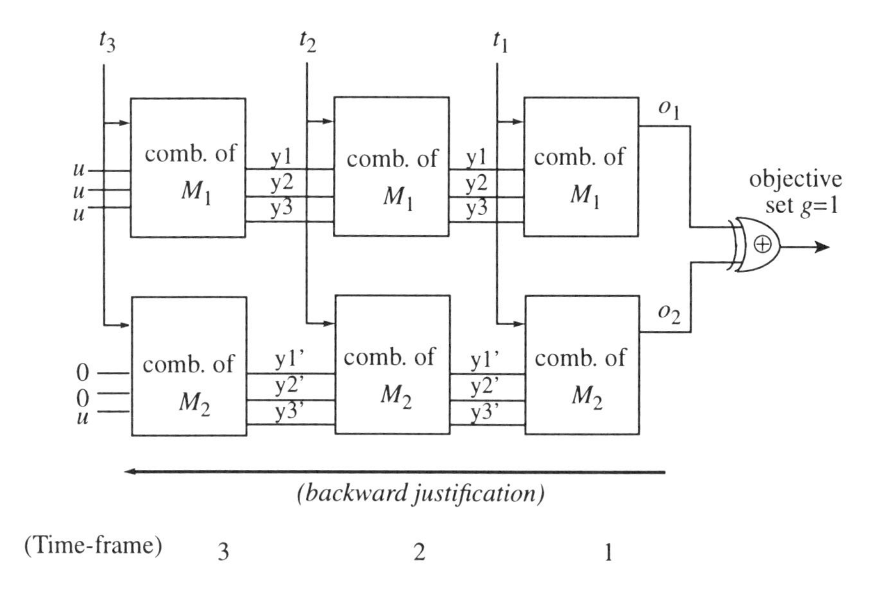

Fig. 4.4 T = (t_3, t_2, t_1) is a distinguishing sequence that can set signal g to '1' from any state covered by (uuu | $00u$).

should be a state requirement of the miter that covers (s_1 | s_2). To check for the sequential hardware equivalence based on Property 4.1, we first apply an input sequence that initializes the two circuits to a state pair, e.g., (q_1, q_2), and then check if (q_1, q_2) is an equivalent state pair.

It is known that the use of 3-valued logic simulation may result in some loss of information [2]. However, for the case of checking for the reset equivalence, the suggested approach is as accurate as the symbolic approach even though the 3-valued logic is employed. The following lemma states that if there exists a distinguishing sequence with respect to the reset equivalence, then this approach can find one, given sufficient time for the search.

Lemma 4.3 Let T be an input sequence (some bits can be don't cares) that can distinguish M_1 and M_2 from the reset state pair (s_1, s_2). Then there exists an input sequence D, found by the backward justification process discussed above, so that the output responses for the sequence based on the 3-valued logic simulation satisfy: $(o_1(s_1, D), o_2(s_2, D)) \in \{(0, 1), (1, 0)\}$.

Proof: See Section 4.6.

4.2.3 Checking Three-Valued Equivalence

To check the 3-valued equivalence, a modified ATPG is used to search for an input sequence that can produce any one of the following 6 illegal output combinations $\{(0, 1), (1, 0), (u, 0), (u, 1), (0, u), (1, u)\}$. Similar to the case of checking the 3-valued safe replaceability, such a sequence is called a 3-valued distinguishing sequence, and will serve as a counter-example to prove that the two circuits are not 3-valued equivalent. If no such input sequence is found after exhausting the entire input space, then M_1 is 3-valued equivalent to M_2. Fig. 4.5 shows the difference of a 3-valued distinguishing sequence and a test sequence for signal g stuck-at-0 of the miter. The following lemma shows the necessary and sufficient condition for 3-valued equivalence that can be checked by the modified ATPG.

Lemma 4.4 An input sequence T_d is a distinguishing sequence with respect to the 3-valued equivalence if and only if at least one of the following two cases is satisfied:

(1) There exists a state s_1 (either fully or partially specified) such that $(o_1(s_1, T_d), o_2(x, T_d))$ belongs to $\{(0, 1), (1, 0)\}$.

(2) There exists a state s_2 (either fully or partially specified) such that $(o_1(x, T_d), o_2(s_2, T_d))$ belongs to $\{(0, 1), (1, 0)\}$.

Proof: Omitted.

4.3 The Speed-Up Techniques

Directly using an ATPG program for the stuck-at-0 fault at the miter's output will cause long CPU run-time and is impractical. In this section, we discuss the speed-up techniques to improve the efficiency of the search process.

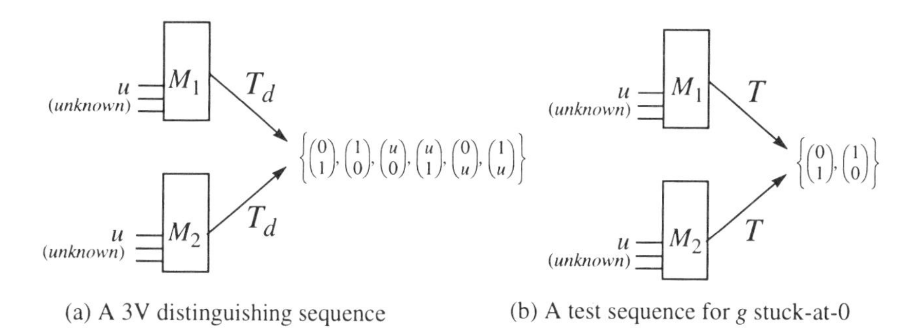

Fig. 4.5 The output combinations of a distinguishing sequence for 3-valued equivalence and a test sequence for miter's output g stuck-at-0 fault.

4.3.1 Test Generation with Breadth-First-Search

A sequential ATPG-program is designed to find a test sequence for a target fault as soon as possible. Hence it usually incorporates a guided depth-first-search strategy. This kind of search strategy is not suitable for the purpose of verifying equivalence. The entire search space has to be explored to conclude a miter's output stuck-at-0 fault is not partially testable. The guidance based on controllability that tries to predict the best decision used in a conventional ATPG program is no longer effective in this application. What could boost the efficiency is the guidance that could trim down the search space. The backward justification process can be characterized as a tree shown in Fig. 4.6.

The root node corresponds to the objective of setting a miter's output to '1'. Each non-root node represents a state requirement to be further justified. A leaf node is a state requirement either not explored yet or proven unjustifiable (denoted by a connection to the ground). We modify the main flow of the test generation process as follows.

(1) *Select an unjustified node* (a leaf node in the tree).

(2) *Perform the pre-image computation for one time frame* to explore all

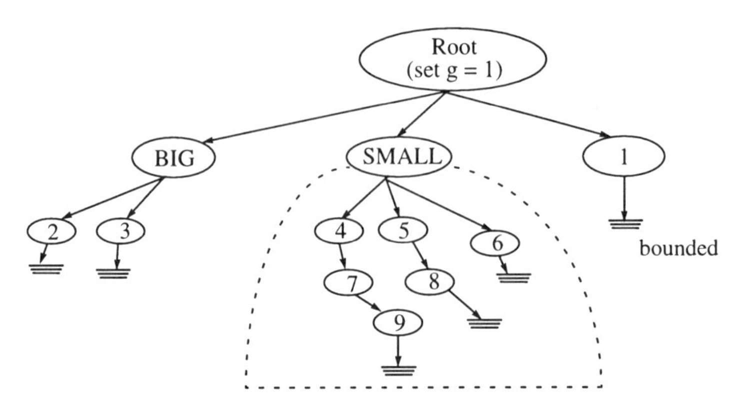

Fig. 4.6 Justification tree.

the children of the target node. The pre-image computation finds all minimally specified cubes in terms of the PI's and the present state lines (PS's) that can satisfy the state requirement at the next state lines. If no pre-image exists, then the selected state requirement is declared as unjustifiable.

(3) *Check the stopping criteria*: if the objective has been justified (a test or partial test has been found) or every node has been explored and proven unjustifiable, then stop. Otherwise, go to step (1) and continue.

This breadth-first search strategy allows a higher degree of flexibility in selecting the next target state requirement for justification than a traditional guided depth-first search strategy. The following property shows the motivation for a heuristic order of selecting the next target state requirement at each iteration.

Property 4.4 (Motivation for largest-first scheduling): Let *BIG* and *SMALL* be two state requirements (3-valued states). If *BIG* covers *SMALL* and BIG is unjustifiable, then *SMALL* is also unjustifiable.

A state requirement can be regarded as a cube in terms of the state variables. The size of the cube can be represented by the number of unknown bits. Property 4.4 suggests that among a set of state requirements to be chosen as the next target for justification, the largest one is a good candidate. If the largest one is justifiable, a distinguishing sequence is found and we can conclude that M_2 is not a 3-valued safe replacement for M_1. If the largest one turns out to be unjustifiable, then any smaller state requirement covered by this candidate is also unjustifiable. In either case, the smaller state requirement can be dropped from further consideration without losing accuracy. This largest-first strategy policy is particularly beneficial when the state requirement *SMALL* has a large justification sub-tree to be traversed to prove it unjustifiable. Fig. 4.6 also illustrates an example that if *BIG* covers *SMALL*, then the sub-tree originated from the node *SMALL* need not be explored.

4.3.2 Exploring the Structural Similarity

Even for a medium-sized circuit, the backward justification routine could be very CPU-time consuming because much effort is wasted in trying to justify the unjustifiable states repeatedly. Similar to combinational verification, the structural similarity between circuits can be explored to improve the efficiency of the backward justification process.

In this generalization, whenever an equivalent internal signal pair is found, a *constraint* is imposed on the two signals *implicitly*, instead of merging them *explicitly* [13]. The constraint is general and could be either an equivalent or covering relation. The advantage of not merging two sequentially equivalent signals explicitly is the following. We are dealing with 3-valued logic instead of binary-valued logic. A covering relation between signals is used when checking the 3-valued safe replaceability and it does not imply the *binary equivalence*. For instance, if internal signal s_2 in M_2 is a safe replacement for signal s_1 in M_1, then the value combinations at (s_1, s_2) belongs to the set $\{(0, 0), (1, 1), (u, u), (u, 0), (u, 1)\}$. If we merge these two signals, then the possibility of the value combinations $\{(u, 0), (u, 1)\}$ would be excluded, and the signal s_1 will be over-specified. Therefore, a signal pair cannot be merged if they satisfy only the covering relation. However, the imposed constraint still provides an effective *bounding condition* during the backward justification process. Whenever the value assignment at a signal pair violates the imposed constraints, backtracking is performed immediately to avoid unnecessary search.

4.3.3 Identifying Equivalent Flip-Flop Pairs

The most important issue in extending the idea of incremental verification for sequential circuits is the identification of the equivalent flip-flop pairs (FF-pairs). A naive way of doing this is to treat each present-state line pair as a pseudo output pair and run the modified ATPG. In our experience, this method is very inefficient. In this subsection, we discuss an *induction-based procedure* to identify the equivalent FF-pairs.

For each candidate equivalent flip-flop pair to be verified for equivalence, the next (present) state line pair of it is referred to as a *candidate NS-pair* (*PS-pair*). We identify the true equivalent FF-pairs by successively removing false FF-pairs from the initial candidate list until no false equivalent FF-pair exists. Thus, it is an iterative screening process.

At each iteration of this monotone screening procedure, *we assume each candidate PS-pair in the candidate list is equivalent and impose an equivalence constraint on each candidate PS-pair.* Then, we run the modified ATPG to verify if each candidate NS-pair is indeed equivalent under these constraints. If all candidate NS-pairs are proven equivalent under the assumption that every candidate PS-pair is equivalent, then we conclude that all candidate FF-pairs remaining in the list are indeed equivalent. On the other hand, if any NS-pair is found not equivalent, then the corresponding flip-flop pair is removed from the candidate list and the process starts over again with a smaller set of candidate equivalent FF-pairs.

The efficiency of this procedure relies on the equivalence constraint imposed on every candidate PS-pair. These constraints provide two advantages: (1) The number of state requirements generated during the backward justification process is reduced substantially, so that our breadth-first search strategy is less vulnerable to memory explosion. (2) The time complexity is substantially reduced because many unjustifiable state requirements that violate the imposed constraints are detected early and, thus, unnecessary search is avoided.

Fig. 4.7 shows an example with two flip-flops in each circuit. Suppose both FF-pairs are in the candidate list initially. Their corresponding PS-pairs are (y_1, y_1') and (y_2, y_2'), and their corresponding NS-pairs are (Y_1, Y_1') and (Y_2, Y_2'). The equivalence of the first NS-pair, (Y_1, Y_1'), is first verified under the constraint that $y_1 = y_1'$, $y_2 = y_2'$ at every time frame. This is done by connecting Y_1 and Y_1' to an exclusive OR gate and running modified ATPG for this gate's output stuck-at-0 fault. Suppose after backward justification for two time frames, only one state requirement, (01 | 00), is left as shown in Fig. 4.7. This state requirement violates one of

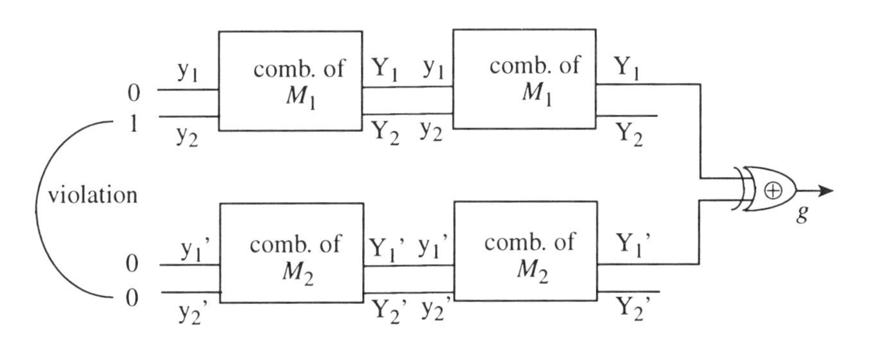

Fig. 4.7 Example for the inductive algorithm to identify equivalent FF-pairs.

the imposed constraints, $y_2 = y_2'$, and thus, is immediately declared as unjustifiable. As a result, (Y_1, Y_1') is regarded as equivalent at this iteration. On the contrary, if (Y_1, Y_1') is proven non-equivalent, then it will be dropped from the candidate list. It can be proved that, *at each iteration of this screening process, the FF-pairs corresponding to the identified non-equivalent NS-pairs are indeed non-equivalent*. Note that the candidate NS-pairs that remain in the list before the end of the process may not be truly equivalent. The number of candidate NS-pairs decreases monotonically until a stable condition is reached, i.e., *every NS-pair in the current candidate list is proven equivalent under the equivalence constraints imposed on the candidate-PS pairs.*

Lemma 4.5 All FF-pairs remaining in the candidate list are indeed equivalent after the stable condition is reached.

Proof: We use induction to prove that all candidate NS-pairs surviving the screening process are equivalent at all time frames. The index of a time frame is increased in a forward manner.

(I. BASIS): Initially, all flip-flops are 'u' and, thus, all candidate FF-pairs are equivalent.

(II. INDUCTION-STEP): Assume that all candidate NS-pairs are equivalent for the first n time frames. This implies that all candidate-PS pairs for the first n+1 time frames are equivalent. Next, it needs to be proved that if the stable condition is reached, then every candidate NS-pair at time frame n+1 is also equivalent.

Suppose there exists an input sequence, T_d, which differentiates at least one candidate NS-pair at time frame n+1, while satisfying the imposed constraints on every candidate PS-pair (i.e., the responses of the two PS-lines in each candidate PS-pair are equivalent). Then, T_d is a distinguishing sequence for at least one candidate NS-pair without violating the imposed constraints at each candidate PS-pair. Therefore, the monotone screening process has not reached the stable condition yet (i.e., at least one candidate NS-pair is false). By contraposition, we conclude that if the stable condition is reached, then every candidate NS-pair at time frame n+1 must be equivalent.

(III. CONCLUSION) Based on the induction, we conclude that all candidate NS-pairs are equivalent *at all time frames* after the stable condition is reached. (Q.E.D.)

The above procedure can be further enhanced by identifying internal equivalent signal pairs. At each iteration, the candidate equivalent internal signal pairs are sorted in an order such that each signal is placed after its transitive fanins. Then, the equivalence of each candidate internal pair is examined based on this order. Once an equivalent pair is found, the equivalence constraint is imposed on that pair to assist the subsequent process of identifying other equivalent pairs. Note that it may take more than one time frame to prove the equivalence of an internal signal pair as in the case of proving equivalence of primary output pairs. We summarize the overall flow in Fig. 4.8.

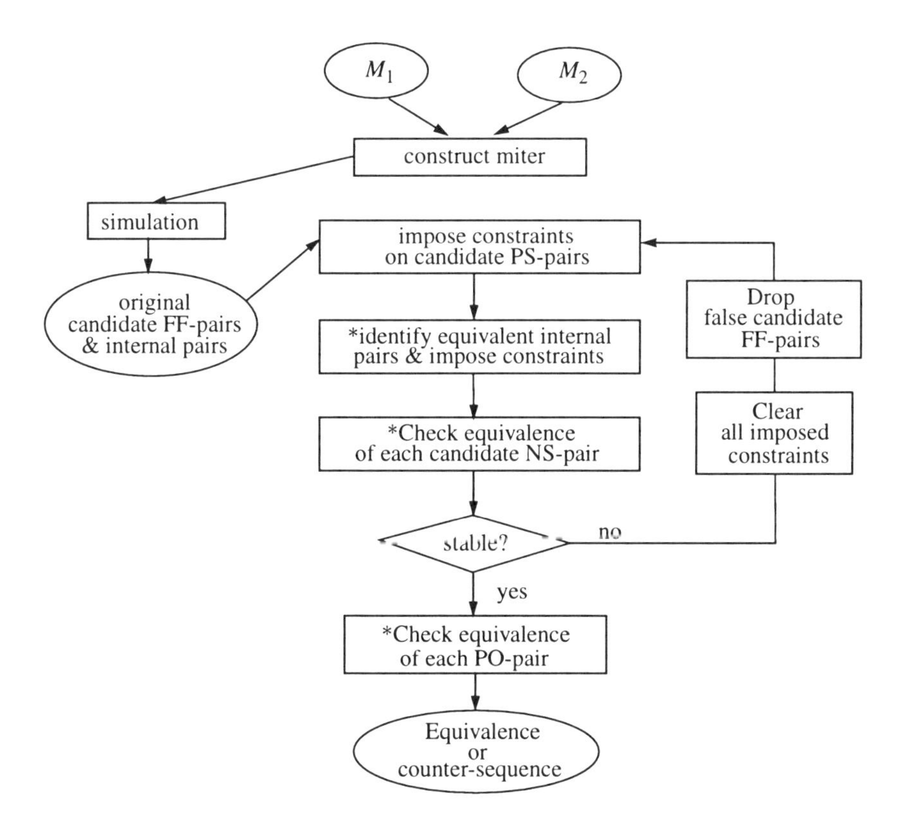

Fig. 4.8 The Overall verification flow based on the ATPG techniques.
(* indicates operations that need to run the modified ATPG)

4.4 Experimental Results

The algorithms discussed in this chapter have been implemented in an ATPG framework [52]. The ISCAS-89 benchmark circuits optimized by a sequential redundancy removal program [32] are verified by this sequential verifier. In Table 4.1, we show the results of checking the 3-valued safe

Table 4.1 Verifying the 3-valued safe replaceability for circuits after sequential redundancy removal.

Circuit	# wires original / optimized	# FFs orig. / opt.	# equivalent FF-pairs (comb. / seq.)	CPU-time (seconds)
s27	17 / 17	3 / 3	3 (3 / 0)	0.1
s298	136 / 118	14 / 14	14 (13 / 1)	4.0
s344	184 / 161	15 / 15	15 (15 / 0)	5.5
s349	185 / 161	15 / 15	15 (15 / 0)	5.4
s382	182 / 163	21 / 21	21 (15 / 6)	5.8
s386	172 / 139	6 / 6	6 (6 / 0)	9.9
s400	186 / 162	21 / 21	20 (12 / 8)	6.3
s444	205 / 173	21 / 21	20 (19 / 1)	6.6
s510	236 / 231	6 / 6	6 (6 / 0)	22.4
s526	217 / 198	21 / 21	21 (20 / 1)	7.4
s713	447 / 201	19 / 19	19 (19 / 0)	13.3
s820	312 / 304	5 / 5	5 (5 / 0)	38.7
s832	310 / 304	5 / 5	5 (5 / 0)	43.5
s1196	561 / 504	18 / 18	16 (16 / 0)	159.3
s1238	540 / 499	18 / 18	16 (16 / 0)	158.0
s1423	748 / 695	74 / 74	74 (74 / 0)	33.1
s1488	667 / 601	6 / 6	6 (6 / 0)	106.9
s1494	661 / 599	6 / 6	6 (6 / 0)	119.2
s5378	3890 / 3390	164 / 137	137 (108 / 29)	460.4
s9234.1	6744 / 6372	211 / 200	189 (188 / 1)	2660.5

replaceability. All optimized circuits are proven as 3-valued safe replacement for their original versions. The CPU times are shown in the last column. Table 4.2, shows the results of verifying the reset equivalence. For circuit s5378, because the optimized circuit is not reset equivalent to the original version, an initialization sequence is first applied to bring both circuits under verification to a state pair and then prove if this state pair is equivalent. Therefore, we examine the sequential hardware equivalence for this case. Every optimized circuit is proven to be reset equivalent or sequential hardware equivalent (for s5378) to their respective original version.

Table 4.2 Verifying the reset-equivalence for circuits after sequential redundancy removal.

Circuit	# FFs orig. / opt.	# equivalent FF-pairs (comb. / seq.)	SIS Time (seconds)	Our Time (seconds)
s27	3 / 3	3 (3 / 0)	0.3	0.5
s298	14 / 14	14 (13 / 1)	0.5	3.7
s344	15 / 15	15 (15 / 0)	8.0	5.1
s349	15 / 15	15 (15 / 0)	8.3	5.2
s382	21 / 21	21 (15 / 6)	51.0	5.7
s386	6 / 6	6 (6 / 0)	1.8	9.5
s400	21 / 21	20 (12 / 8)	49.9	6.1
s444	21 / 21	20 (19 / 1)	37.5	6.4
s510	6 / 6	6 (6 / 0)	3.6	21.1
s526	21 / 21	21 (20 / 1)	0.9	7.2
s713	19 / 19	19 (19 / 0)	12.4	12.8
s820	5 / 5	5 (5 / 0)	5.6	37.1
s832	5 / 5	5 (5 / 0)	5.8	44.5
s1196	18 / 18	16 (16 / 0)	13.1	160.0
s1238	18 / 18	16 (16 / 0)	13.1	152.0
s1423	74 / 74	74 (74 / 0)	—	32.8
s1488	6 / 6	6 (6 / 0)	8.4	105.8
s1494	6 / 6	6 (6 / 0)	9.0	117.5
s5378	164 / 137	137 (108 / 29)	—	456.9
s9234.1	211 / 200	198 (188 /10)	—	1471.4

The CPU times of running the verification program in SIS [121] (based on the symbolic approach) are given in the last second column. The new program successfully verifies three larger circuits (s1423, s5378, s9234.1) for which SIS fails on a Sun Sparc-5 workstation with 128-M byte memory. The new approach is efficient due to several techniques used to cut down the search space. Among them, the most powerful technique is the one for identifying the equivalent FF-pairs (the number of identified equivalent FF-pairs is shown in the column labeled *# equivalent FF-pairs*). Due to these techniques, the number of state requirements generated in the ATPG search process is significantly reduced and the possibility of running into memory explosion is dramatically reduced. The table also lists the numbers of combinationally equivalent FF-pairs and the sequentially equivalent FF-pairs identified in our program. The difference between these two is that the latter requires more than one time frame in ATPG to identify their equivalence. For circuit s9234.1, the number of identified equivalent FF-pairs in checking the reset-equivalence (198) is different from the number in checking the 3-valued safe replaceability (189). As a consequence, checking the reset equivalence is faster than checking the 3-valued safe replaceability for this circuit.

4.5 Summary

This chapter introduces an ATPG-based framework that can verify the equivalence of two sequential circuits with or without a reset state. We prove that this approach is correct and complete for circuits with a reset state. For circuits without a reset state, this approach can verify the design revision with respect to either the sequential hardware equivalence, the 3-valued safe replaceability, or the 3-valued equivalence. We show that the 3-valued safe replaceability is a more stringent definition than the sequential hardware equivalence for 3-valued initializable circuits. This approach is particularly effective in the cases where two circuits possess significant structural similarity, e.g., circuits optimized by sequential redundancy removal and/or by retiming, and circuits that have gone through an engineering change.

We also described several techniques to speed up the verification process: (1) a breadth-first-search with largest-first scheduling to cut down the search space, (2) exploration of the internal similarity to perform the verification in stages, and (3) an iterative procedure to identify the equivalent FF-pairs. Experimental results on verifying circuits after sequential redundancy removal with up to several hundred flip-flops demonstrate that this approach has potential to handle larger designs than the traditional symbolic approach.

4.6 Appendix

In this section, we prove Lemma 4.2 and Lemma 4.3. Lemma 4.2 is the necessary and sufficient condition for the *3-valued safe replaceability*. Lemma 4.3 is about the correctness of the new method when applied to circuits with a reset state. In the following, we first introduce the enhanced 3-valued logic simulation.

Definition 4.10 (*enhanced 3-valued logic simulation*) Let v be an input vector. The value of a signal f in response to the input vector v, denoted as $f(v)$ is determined by one of the three cases:

(1) If $f(w) = 0$ for every binary vector w covered by v, then $f(v) = 0$.

(2) If $f(w) = 1$ for every binary vector w covered by v, then $f(v) = 1$.

(3) If there exist two binary vectors, w_1 and w_2, covered by v such that $f(w_1) = 0$ and $f(w_2) = 1$, then $f(V) = u$.

Lemma 4.2: M_2 is a 3-valued safe replacement for M_1 if and only if there exists no input sequence T and state s_2 of M_2 such that $(o_1(x, T), o_2(s_2, T))$ is either $(0, 1)$ or $(1, 0)$.

Proof: (I) ($\Rightarrow$) Assume that there exists an input sequence T such that $(o_1(x, T), o_2(s_2, T))$ is either $(0, 1)$ or $(1, 0)$. Then the possible output combinations of signal o_1 and o_2 in response to input sequence T (when both circuits started from the unknown state) belongs to $\{(0, 1), (1, 0), (0, u), (1, u)\}$. Thus T is a distinguishing sequence for M_1 and M_2. By contraposition, we conclude that if M_2 is a safe replacement for M_1, then there exists no input sequence T and a state s_2 such that $(o_1(x, T), o_2(s_2, T))$ belongs to $\{(0, 1), (1, 0)\}$.

(II) ($\Leftarrow$): Assume that M_2 is *not* a safe replacement of M_1. Then there exists a distinguishing sequence such that the output combination $(o_1(x, T), o_2(x, T))$ belongs to $\{(0, 1), (1, 0), (0, u), (1, u)\}$ using the enhanced 3-valued logic simulation.

We assume that T is a *true* distinguishing sequence in the sense that if T produces $(0, u)$ or $(1, u)$ at some primary output pair, the logic value u is a true u, and is not due to the limitation of conservative 3-valued logic simulation. For example, if the output combination $(o_1(x, A), o_2(x, A))$ is $(0, u)$ using the traditional 3-valued logic simulation, and $(o_1(x, A), o_2(x, A))$ is $(0, 0)$ using the enhanced 3-valued logic simulation. Then, we consider A to be an *ambiguous distinguishing sequence*, and not a true distinguishing sequence.

Based on this assumption, if T produces $(0, u)$ or $(1, u)$ at some primary output pair, there exists a binary state q_2 such that $(o_1(x, T), o_2(q_2, T))$ belongs to $\{(0, 1), (1, 0)\}$ using not only the enhanced 3-valued logic simulation but also the traditional 3-valued logic simulation. We thus conclude by contraposition that if there exists no input sequence T such that $(o_1(x, T), o_2(s_2, T))$ belongs to $\{(0, 1), (1, 0)\}$, then M_2 is a safe replacement for M_1. (Q.E.D.)

Consider the scenario of verifying a design consisting of a number of sub-networks. If every transformed sub-network is a 3-valued safe replacement for its original, then under the assumption that some *ambiguous* distinguishing sequences may still exist, it can be proven that every initialization sequence for the original design still *synchronizes* (although may not initialize) the transformed design. In addition, the transformed design is equivalent to the original one with respect to SHE. The reason is that every input sequence producing a deterministic '0' or '1' in the original design using traditional 3-valued logic simulation will still produce the same '0' or '1' in the transformed design using the enhanced 3-valued logic simulation.

Lemma 4.3 Let T be an input sequence (some bits could be u) that can distinguish M_1 and M_2 from the reset state pair (s_1, s_2). Then there exists an input sequence T_d that can be found during the backward justification process in the algorithm described in Section 4.2 such that $(o_1(s_1, T_d), o_2(s_2, T_d))$ is either (0, 1) or (1, 0) using the traditional 3-valued logic simulation.

Proof: We construct a binary input sequence T' from T by replacing every unknown bit with the logic value '0'. Obviously, T' is a refinement of T and is still a distinguishing sequence. Let $T' = (t_1, t_2,..., t_n)$, where t_i is a binary vector. Assume that T will bring the miter through a sequence of completely specified states $(q_1, q_2,..., q_n, q_{n+1})$ from the initial state q_1 as shown in Fig. 4.9, where $q_1 = (s_1 \mid s_2)$.

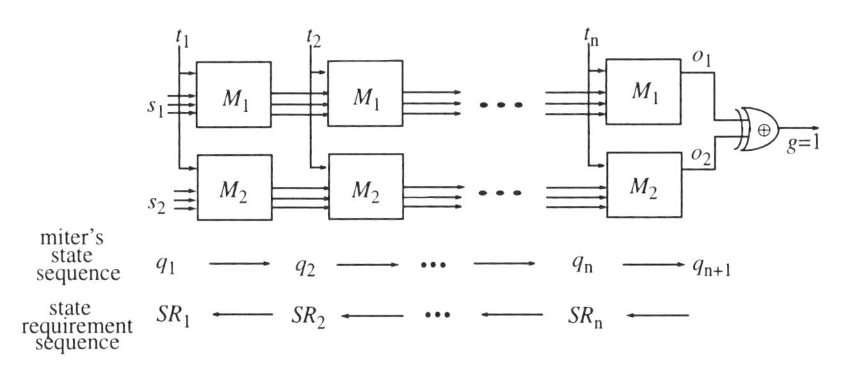

Fig. 4.9 Time-frame expansion model for proving Lemma 4.3.

We prove by induction that there exists an input sequence $T_d = (d_1, d_2,..., d_n)$ that can be found during the backward justification process.

(BASE): At the right-most time-frame, since the binary vector $(t_n \mid q_n)$ can set signal g to 1, a state requirement SR_n and an input vector d_n can be found during the backward justification process of this time-frame so that the following two conditions are satisfied: (1) $(d_n \mid SR_n)$ can set g to 1, and (2) $SR_n \supset q_n$. Otherwise the justification of this time frame is not complete.

It is known that the combinational backward justification is a complete process even though the over-specification exists at some signals.

(INDUCTION-STEP): Suppose at the i-th time frame, there exists a state requirement SR_i such that $SR_i \supset q_i$. Similar to the above argument, we can find an input vector d_{i-1} and a state requirement SR_{i-1} such that $SR_{i-1} \supset q_{i-1}$ and $d_{i-1} \supset t_{i-1}$.

(CONCLUSION): Given sufficient time, a state requirement that covers the reset state $(s_1 \mid s_2)$ will be generated at the first time frame. This state requirement satisfies the justified condition of the backward justification process. The input vectors collected so far is a 3-valued distinguishing sequence. (Q.E.D.)

Chapter 5

AQUILA: A Local BDD-based Equivalence Verifier

In this chapter we describe the enhancement of the ATPG-based framework by local BDD-based techniques. This enhancement involves two ideas. First, we generalize the inductive algorithm of Section 4.3 to identify equivalent flip-flop pairs and sequentially equivalent internal signal pairs. Secondly, we incorporate a heuristic called *partial justification* to handle larger designs using local BDDs. The approach is much less vulnerable to a memory explosion than the traditional symbolic FSM traversal and is, therefore, suitable for real-life designs. A prototype tool called AQUILA is implemented to demonstrate the efficiency and advantage of this method.

5.1 Overall Flow

For simplicity, we assume that both circuits under verification have an external reset state. The discussion can be generalized for circuits without a reset state based on the definitions of sequential hardware equivalence or 3-valued safe replaceability. Fig. 5.1 shows the overall flow of a sequential gate-to-gate equivalence checker that incorporates a hybrid approach

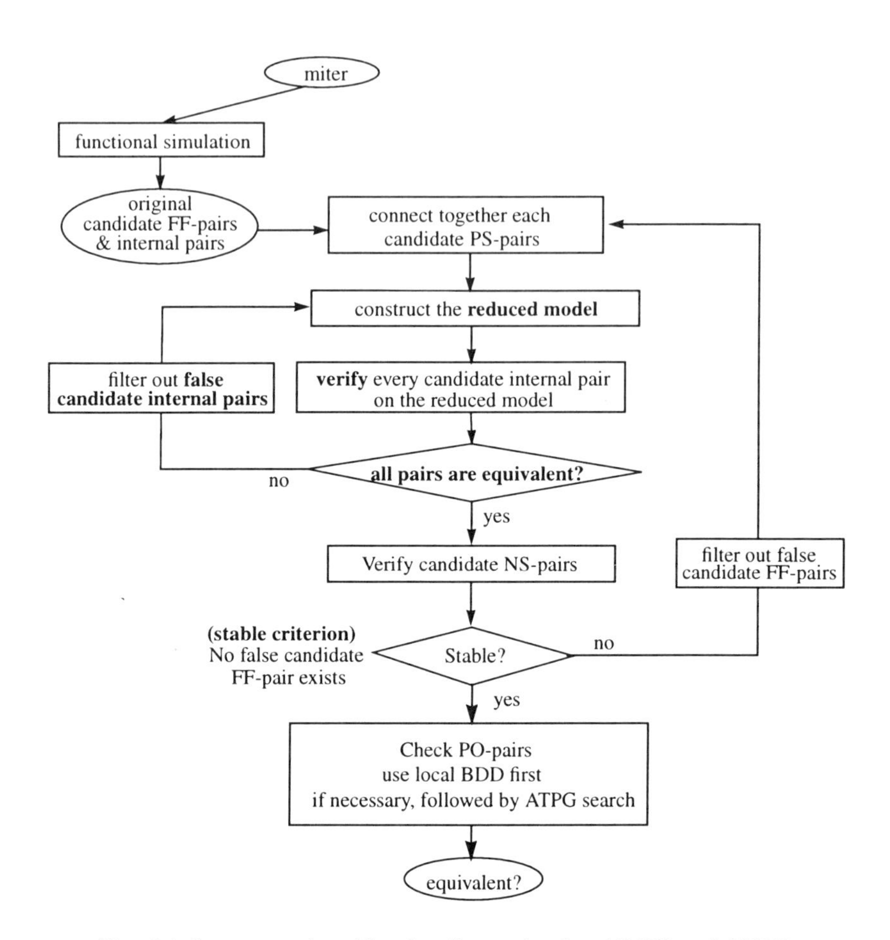

Fig. 5.1 Incremental verification flow using local BDD and ATPG.

employing both ATPG and local BDD techniques. In general, local BDDs are more efficient in proving the equivalence of an internal signal pair or a primary output pair, while ATPG is more effective in finding a counter-example if one exists. Combining the advantages of these two techniques makes this approach capable of handling many practical situations. There are primarily three phases in this algorithm:

Phase 1: Construct the miter and run simulation using a number of functional vectors to find the candidate equivalent flip-flop pairs and candidate internal signal pairs.

Phase 2: Simplify the miter in stages by identifying true equivalent flip-flop pairs and equivalent internal signals by a generalized inductive algorithm. This generalized algorithm consists of a two-level loop. The outer loop identifies equivalent flip-flop pairs and the inner loop identifies equivalent internal signal pairs. The details of this algorithm are discussed in the next section.

Phase 3: Check the equivalence of each primary output pair using a local BDD-based technique, followed by an ATPG-based technique, if necessary.

The rest of this chapter is organized as follows. Section 5.2 discusses the ideas of exploring sequential similarity between the two CUVs by a generalized inductive algorithm. Section 5.3 discusses the computational details of checking the sequential equivalence of a signal pair using symbolic backward justification and the heuristic of partial justification. Section 5.4 presents experimental results, and Section 5.5 concludes the chapter.

5.2 Two-Level Inductive Algorithm

The inductive algorithm of the previous chapter uses the concept of *assume-and-then-verify* to identify the equivalent flip-flop pairs. This algorithm allows equivalence checking to be performed on a simplified miter, instead of the original miter, and thus, can efficiently identify equivalent flip-flop pairs. For the cases when the design has a reset state, Fig. 5.2 shows an example of the simplified model where $\{(y_1, z_1), (y_2, z_2)\}$ are regarded as the candidate flip-flop pairs. As opposed to the algorithm described in Section 4.3, where we only impose the equivalence constraint on each candidate PS-pair, here we actually merge them (or more precisely, replace y_1 and y_2 by z_1 and z_2 respectively). We call this model as the *first-level reduced model*.

It is worth mentioning that the efficiency of this algorithm depends on the underlying techniques of equivalence checking. To demonstrate this

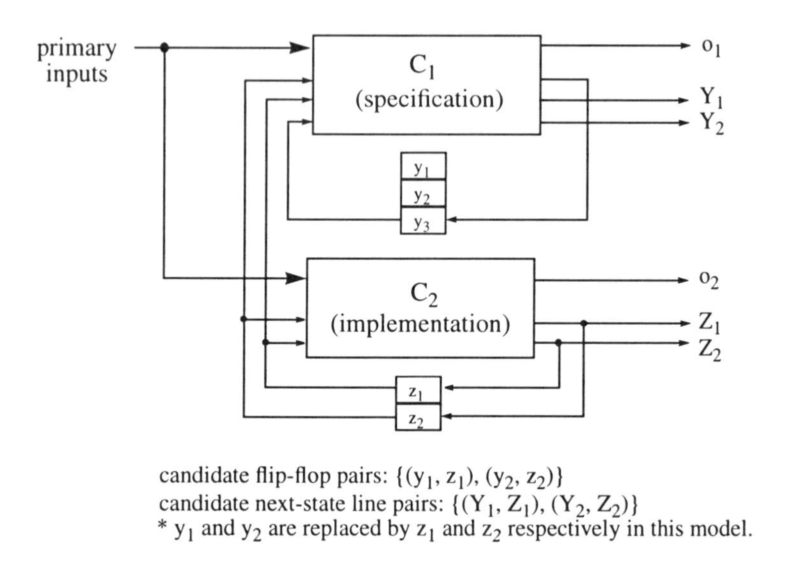

Fig. 5.2 The first-level reduced model for exploring sequential similarity.

point, consider an extreme case in which the two designs have a perfect one-to-one flip-flop correspondence and are indeed combinationally equivalent. In the first-level reduced model, it needs to be proven that each candidate NS-pair is equivalent. If the traditional forward FSM traversal (as described in Section 2.2) is used, then this process still requires the exploration of the entire state space of the design. On the other hand, if the backward justification techniques (as described in Chapter 4) is employed, then only one time frame is sufficient to prove the equivalence for each candidate NS-pair. The difference in computational complexity between these two is huge. There exist many cases where state traversal cannot be completed, while the BDD-representation of the combinational portion can be easily constructed. This example indicates that the advantage of this speed-up technique can be fully exploited only if it is accompanied with the backward justification techniques.

5.2.1 Second-Level Assume-And-Then-Verify

The process of proving the equivalence of a candidate flip-flop pair can be further speeded up by identifying equivalent internal signal pairs in the first-level reduced model. Consider the example shown in Fig. 5.3(a). In addition to the candidate flip-flop pairs, suppose we have three candidate internal pairs, $\{(a, a'), (b, b'), (c, c')\}$, derived from the results of the preprocessing stage.

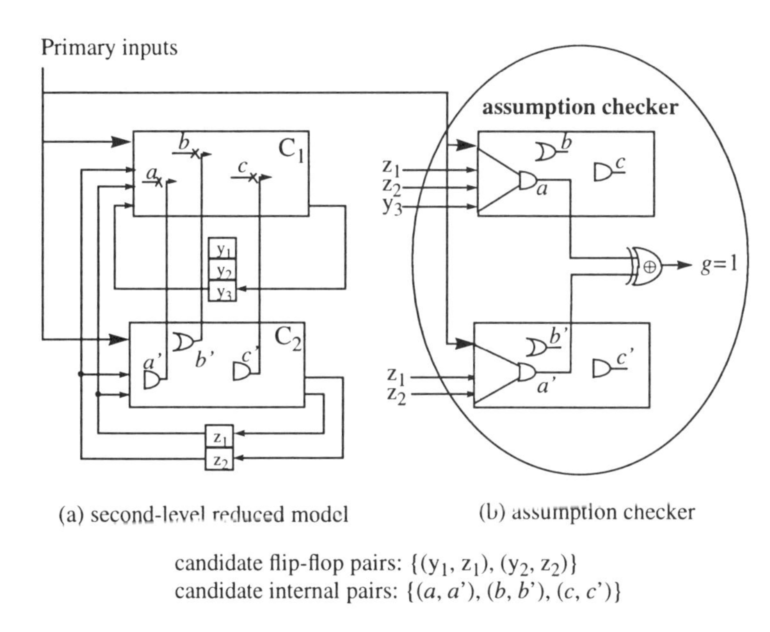

Fig. 5.3 Second-level reduced model and the assumption checker.

Similar to the identification of equivalent flip-flop pairs, we incorporate a monotone filter in this generalization. Initially, we assume that all internal signal pairs are equivalent. Then, we iteratively identify false candidate internal pairs (as will be defined later) and remove them from the candidate list. This process continues until no false candidate pair exists.

Identifying false candidate pairs is an iterative process as well. Each iteration involves two steps:

Step 1: Construct a *second-level reduced model* that merges the candidate PS-pairs as well as the candidate internal pairs (Fig. 5.3(a)). Also, a copy of the combinational portion of the first-level reduced model, called *assumption checker*, is created and attached to the reduced model as shown in Fig. 5.3(b).

Step 2: Check the equivalence of each candidate internal pair in the assumption checker in stages. A candidate pair (a_1, a_2) is *false* if there exists an input sequence that differentiates (a_1, a_2) in the assumption checker network. Fig. 5.3(b) shows a snapshot during the checking of the equivalence of signal pair (a_1, a_2). The computational details will be described in the next section. If no false candidate pair exists, then the process reaches a stable condition and exit. Otherwise, the process starts over with a smaller set of candidate internal pairs after excluding the identified false candidate pairs.

When the iterative process reaches a stable condition, it can be proved by induction that each survivor candidate pair is indeed equivalent.

Property 5.1 If an input sequence applied to the first-level reduced model does not differentiate any candidate internal pair, then it will bring the first-level and second-level reduced models to the same final state.

Lemma 5.1 Every candidate internal pair remaining in the candidate list is indeed equivalent in the first-level reduced model after the stable condition of the above two-step process is reached.

Proof: We prove that the survivor candidate pair is equivalent for any input sequence by induction. The induction is on the length of the input sequence.

(I. BASIS): For any input sequence of length '1', every candidate pair is equivalent. Otherwise, some candidate pairs in the assumption checker network would have been classified as false pairs.

(II. INDUCTION-STEP): Assume that every candidate internal pair is equivalent for any input sequence of length n in the first-level model. From

Property 5.1, it follows that the set of reachable states of the first-level and second-level reduced models in n clock cycles are the same. As a result, if we cannot differentiate any candidate pair in the assumption checker network of the second-level reduced model in n+1 cycles, then we cannot differentiate any pair in the first-level reduced model either. Therefore, every candidate internal pair is equivalent for any input sequence of length n+1 once the stable condition is satisfied.

(III. CONCLUSION) Based on the above inductive analysis, we conclude that every candidate internal pair that survives the filtering process is indeed equivalent *for any input sequence* in the first-level reduced model. (Q.E.D.)

The above lemma can also be interpreted as, *if there exists at least one false pair, then the suggested algorithm will identify at least one pair* (*i.e., the process will not be stabilized*). Fig. 5.4 shows the pseudo-code of AQUILA.

5.3 Symbolic Backward Justification

In this subsection, we describe the symbolic technique to check the sequential equivalence of a signal pair (a_1, a_2) in the assumption checker using the local BDD. Computationally, this is performed on the iterative array model. Contrary to an ATPG-based procedure, the backward justification is now performed by a sequence of symbolic pre-image computations.

At the last time frame of this procedure, the two signals in the target pair (a_1, a_2) are treated as pseudo primary outputs and the set of input vectors at primary inputs (PI's) and present state lines (PS's) that can differentiate this target pair is computed. Fig. 5.5 shows an illustration. The characteristic function of this set is called the discrepancy function at time frame 0, denoted as $Disc^0(a_1, a_2)$. In the rest of this chapter, the superscript of a notation indicates the index of the associated time frame. The index of the last time frame is 0 and increases in a backward manner as shown in Fig.

```
AQUILA (M1, M2, Pf, Pi)
        Pf = candidate equivalent flip-flop pairs;
        Pi = candidate internal pairs sorted from PI to PO;
{
    Let A = assumed sequentially equivalent signal pairs;
        Ff = false flip-flop pairs;
        Fi = false candidate internal pairs;

    while(1) { // first level loop
        construct the first-level reduced model;

        /*--- Step 1: identify equivalent internal signal pairs ---*/
        while(1){ // second level loop
            /*--- simplify miter and decide candidate pairs to be assumed ---*/
            set A = φ;
            foreach internal pair p = (a1, a2) in Pi {
            check if equivalence for a single vector by local BDD;
            if (equivalent) replace a1 by a2;
            else if (this pair is critical according to a heuristic) {
            A = A + p;
            replace a1 by a2; // assume sequentially equivalent
            }
            }
            /*--- check assumption ---*/
            construct the 2nd-level reduced model with assumption checker network;
            Fi = φ;
            foreach p = (a1, a2) in A {
            check sequential equivalence of (a1, a2) by partial justification;
            if( ! equivalent ) { Fi = Fi + p; }
            }
            if (Fi != φ) Pi = (Pi - Fi ); // remove false candidate internal pairs
            else break; // no false candidate internal pair, exit 2nd loop;
        }
        /*--- Step 2: check the equivalence of each next-state pair ---*/
        Ff = φ;
        foreach NS-pair p in Pf {
            check equivalence of p;
            if( ! equivalent ) {Ff = Ff + p; }
        }
        /*--- Step 3: check the stable condition ---*/
        if (Ff != φ) Pf = (Pf - Ff ); // remove false candidate flip-flop pairs
        else break; // no false candidate flip-flop pair, exit 1st loop;
    }
    /*--- Step 4: check the equivalence of primary output pair ---*/
    Apply local BDD followed by ATPG search; report any inequivalent PO-pair;
}
```

Fig. 5.4 Pseudo-code of AQUILA.

5.5. After smoothing out all primary inputs of $Disc^0(a_1, a_2)$, we arrive at a new function in terms of the present state lines only. This new function, denoted as $SR^0(a_1, a_2)$, characterizes the set of the state requirements at the last time frame for differentiating (a_1, a_2). Then the checking of the equivalence of (a_1, a_2) is reduced to a sequential backward justification process that decides if there exists an input sequence that will bring the reduced miter from the reset state to any state in $SR^0(a_1, a_2)$.

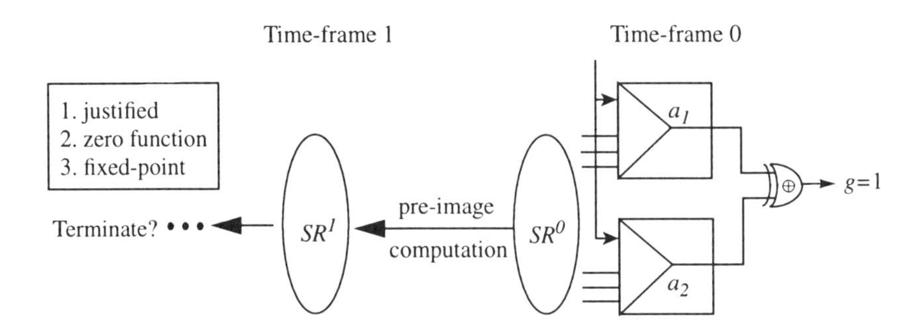

Fig. 5.5 Illustrating a symbolic procedure for checking if $a_1 = a_2$.

A symbolic backward justification process consists of a sequence of perimage computations and smoothing operations. At each time frame i, $(i > 0)$, $SR^i(a_1, a_2)$ is derived by first computing the pre-image of $SR^{(i-1)}(a_1, a_2)$ followed by the smoothing operation that existentially quantifies out every primary input. The symbolic techniques for pre-image computation described in Section 2.2 can be directly applied. The stopping criteria of this backward justification process will be checked after $SR^i(a_1, a_2)$ is derived at time frame i to see if another backward time frame expansion is needed: (1) Justified criterion: when the reset state is contained in $SR^i(a_1, a_2)$. (2) Unjustifiable criterion: when $SR^i(a_1, a_2)$ is a zero function. (3) Fixed-point (or cyclic) criterion: when $SR^i(a_1, a_2)$ does not contain the reset state, but is contained in the union of the state requirements derived so far, $\sum_{k=0}^{i-1} SR^k(a_1, a_2)$. If the unjustifiable criterion or the fixed-point criterion is met, then the target signal pair is an equivalent pair. On the other hand, if the justified criterion is encountered, then it is not an equivalent pair. The pseudo code of this process is provided in Fig. 5.6.

```
Symbolic-equivalence-checking (a1, a2)
{
    compute discrepancy function Disc^0(a1, a2) at the last time frame;
    derive the state requirement function SR^0(a1, a2) by smoothing out PI's;
    New = SR^0(a1, a2);// new state requirements to be further justified
    SR = φ;   // accumulated state requirements so far
    /*------ symbolic backward justification ------*/
    while(1) {
        c = check_stopping_criteria(New);
        switch(c){
            case JUSTIFIED: return(DIFFERENT);
            case UNJUSTIABLE: return(EQUIVALENT);
            case FIXED-POINT: return(EQUIVALENT);
            default:SR = SR ∪ New; // update accumulated state requirements
            New = compute_preimage_and_smooth_out_PI's(SR);
            break;
        }
    }
}
```

Fig. 5.6 Sequential equivalence checking by symbolic backward justification.

5.3.1 Partial Justification

The above symbolic backward justification may still cause a memory explosion for circuits with large number of flip-flops even though it is performed on the reduced model. Hence, a simple heuristic called *partial justification* is further used to enhance the approach. This heuristic can be used during the backward justification process for identifying equivalent flip-flop pairs, internal equivalent pairs, or for checking primary output equivalence.

During the verification process described in Fig. 5.1, assume that the identified equivalent pairs or assumed equivalent pairs in the fanins of (a_1, a_2) of the reduced model have been merged. Now the verification process proceeds to check if a_1 is equivalent to a_2. At the last time frame, instead of computing discrepancy function in terms of the PI's and PS's directly, we select a local cutset, denoted as λ_0, in the pair's fanins. We use $Disc_{\lambda_0}^{0}(a_1, a_2)$ to denote the characteristic function of the set of value combinations at λ_0 that can differentiate (a_1, a_2). The cutset signals in λ_0 should be either previously identified *merge points* (Def. 3.4), or primary inputs. Using the heuristic of *dynamic support* described in Section 3.1.2, expanding the cutset towards the PI's may increase the success rate in proving the equivalence of the target pair. Suppose the cutset has been pushed back towards the PI's and PS's for a number of logic levels within the same time frame but (a_1, a_2) still cannot be proven equivalent, i.e., $Disc_{\lambda_0}^{0}(a_1, a_2)$ is not a zero function. At this point, the cutset λ_0 may contain some present state lines, some primary inputs and some internal signals. We then smooth out the signals in λ_0 that are not present state lines from $Disc_{\lambda_0}^{0}(a_1, a_2)$ to obtain a *superset* of the state requirement function, denoted as $SR_{\lambda_0}^{0}(a_1, a_2)$. This function characterizes a necessary condition that should be satisfied at the present state lines in order to differentiate (a_1, a_2).

Similar to the complete symbolic justification, a number of time frames may need to be explored until one of the three stopping criteria is satisfied. For simplicity, we assume only one cutset is selected at each time frame i, denoted as λ_i. The *over-estimated* state requirement function at time frame i, $SR_{\lambda_i}^{i}(a_1, a_2)$, can be derived from $SR_{\lambda_{i-1}}^{i-1}(a_1, a_2)$ by pre-image computation and smoothing out non-PS supporting signals in λ_i. After this partial justification process, if the over-estimated state requirements for differentiating (a_1, a_2) satisfy the unjustifiable or fixed-point criterion, then it can be proven that (a_1, a_2) is equivalent. However, if the over-estimated state requirements are justified (i.e., reachable from the reset state), the target signal pair may not be indeed inequivalent unless we use only the PI's and PS's as the cutset, in which case the partial justification degenerates to a complete justification.

Selecting a good cutset is essential for this heuristic. Similar to the combinational cases, we expand the cutset dynamically from the target pairs towards primary inputs and present state lines within each time frame. If this heuristic cannot conclude whether or not a pair is equivalent, we take the following action. If the target pair consists of internal signals, we pessimistically assume that they are not equivalent. If the target pair is a flip-flop pair or a primary output pair, then we run ATPG on the target pair to prove the equivalence or to find a distinguishing sequence.

5.3.2 An Example

Fig. 5.7 shows an example. We have four primary inputs x_1, x_2, x_3, and x_4. C_1 and C_2 contain sub-networks, sub_1 and sub_2, respectively. Suppose the outputs of these two sub-networks, s_1 and s_2, have been proven equivalent. In addition to these two sub-networks, C_1 (C_2) has two flip-flops denoted as Z_1 and Z_2 (Y_1 and Y_2), respectively. We use the same label for a logic gate and its output signal. As for flip-flops, the outputs of Z_1 and Z_2 (Y_1 and Y_2) are present state lines and denoted as z_1 and z_2 (y_1 and y_2), respectively. For simplicity, we only apply one-level inductive algorithm in this example. The verification procedure is detailed as follows.

(Step 1): Perform simulation for a large number of random or functional sequences to find the set of candidate flip-flop pairs $\{(Z_1, Y_1), (Z_2, Y_2)\}$, and the set of candidate internal pairs $\{(a_1, a_2), (b_1, b_2), (c_1, c_2), (d_1, d_2), (s_1, s_2)\}$.

(Step 2): Identify equivalent flip-flop pairs.

(2.1) Assume the candidate present state line pairs $\{(z_1, y_1), (z_2, y_2)\}$ are equivalent and connect each candidate PS-pair together.

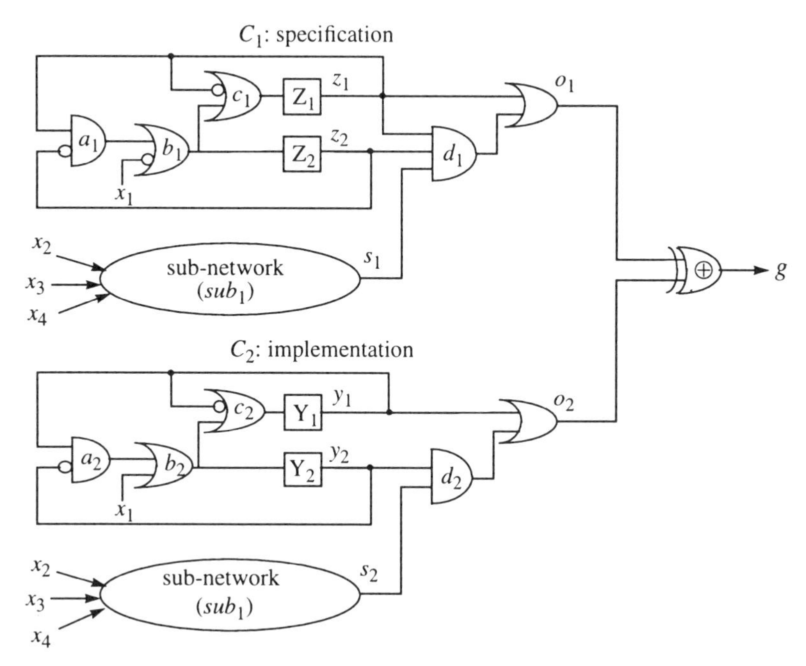

Fig. 5.7 An example for gate-to-gate equivalence checking.

(2.2) Verify the equivalence of $\{(Z_1, Y_1), (Z_2, Y_2)\}$. Sweep the miter from PI's towards PO's to sequentially identify internal equivalent signal pairs. Signal pairs $\{(a_1, a_2), (b_1, b_2), (c_1, c_2)\}$ can be easily verified as equivalent one by one, and so can $\{(Z_1, Y_1), (Z_2, Y_2)\}$. Hence, we conclude that the assumption made in (2.1) is correct (i.e., (Z_1, Y_1) and (Z_2, Y_2) are indeed equivalent) and the process has stabilized.

(Step 3): Check the equivalence of primary output pair (o_1, o_2).

(3.1) Further explore the similarity by checking if (s_1, s_2) and (d_1, d_2) are equivalent pairs. Suppose that (s_1, s_2) is proven equivalent, and thus, we replace s_1 by s_2. The process moves on to check (d_1, d_2). Fig. 5.8 shows a snapshot of the miter at this moment. Suppose we select

$\lambda = \{y_1, y_2, s_2\}$ as the cutset. Then, the distinguishing vector at this

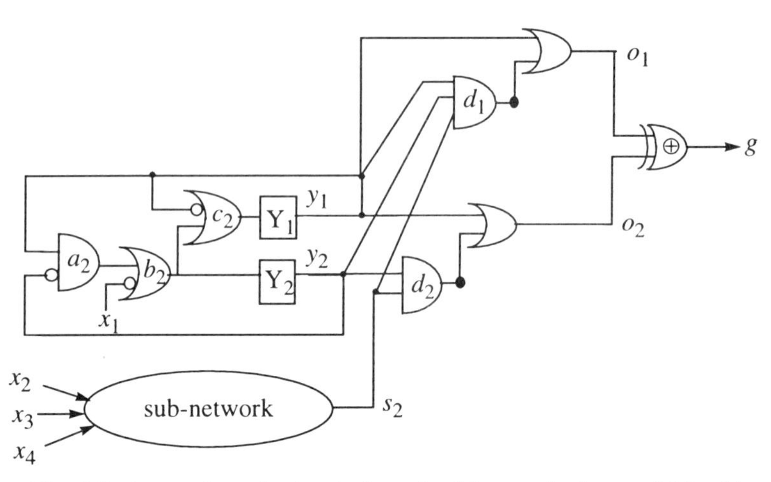

Fig. 5.8 A snapshot of miter before checking equivalence of (d_1, d_2).

cutset is $\{(y_1, y_2, s_2) = (0, 1, 1)\}$ for differentiating d_1 and d_2. Let the characteristic function of this set be $Disc_\lambda^0(d_1, d_2)$. Since signal s_2 is not a present state line, it is smoothed out from $Disc_\lambda^0(d_1, d_2)$ to derive the set of over-estimated state requirements, which is $\{(y_1, y_2) = (0, 1)\}$. None of the three stopping criteria is satisfied. Hence the backward justification process starts. At the next time frame, the pre-image of $\{(Y_1, Y_2) \mid (0, 1)\}$ is an empty set and, thus, the state requirement $\{(y_1, y_2) = (0, 1)\}$ is unjustifiable. Signal pair (d_1, d_2) is sequentially equivalent and the two signals are merged together. The resulting circuit is shown in Fig. 5.9.

(3.2): Check output pair (o_1, o_2) using the local BDD. They can be proved equivalent by selecting $\{y_1, d_2\}$ as the cutset in Fig. 5.9.

(Step 4): Run ATPG on the simplified miter. This is not necessary in this example.

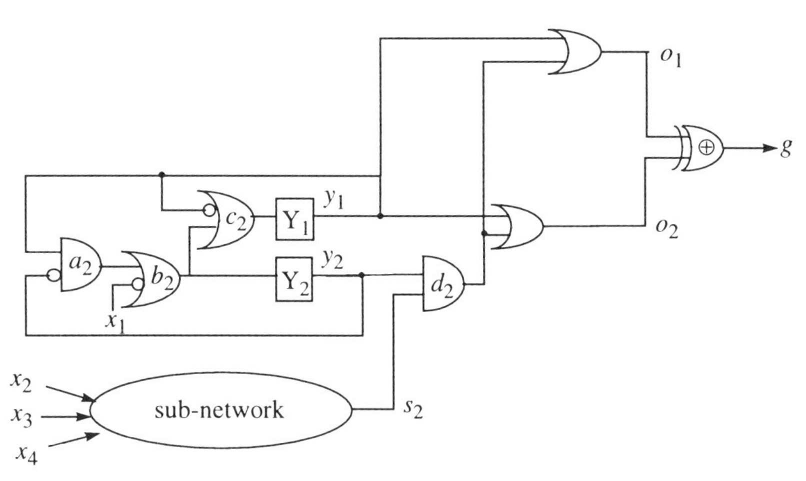

Fig. 5.9 A snapshot of miter after merging d_1 and d_2.

5.4 Experimental Results

We have implemented a prototype tool, called AQUILA, in the SIS environment [121], and integrated a sequential ATPG program *stg*3 [35] in the system. Table 5.1 shows the results of using AQUILA for verifying ISCAS-89 sequential benchmark circuits after they were optimized by a combinational script *script.rugged* in SIS. The optimization process reduces the number of flip-flops for circuits s641, s713, s5378, s13207, and s15850. For these circuits, combinational verification programs cannot be applied even if the inputs (outputs) of the flip-flops are treated as pseudo-outputs (pseudo-inputs). Our program automatically verifies that these optimized circuits are indeed sequentially equivalent to their original versions.

The key columns of Table 5.1 are explained as follows. (1) *nodes (original / optimized)*: the numbers of nodes in the original and the optimized circuits, respectively. The optimized circuits are further cleaned up by SIS command *"sweep"* and then decomposed into AND/OR gates.

(2) *# FFs (orig. / opt.)*: are the numbers of flip-flops in the original and optimized circuits. For instance, s13207 has 490 flip-flops, but only 453 are left in the optimized circuit. (3) *# equiv. internal pairs (comb / seq)*: the number of internal signal pairs that are identified as combinationally equiv-

Table 5.1 Results of verifying ISACS-89 circuits optimized by combinational script *script.rugged* in SIS.

Circuit	# nodes original / optimized	# FFs orig. / opt.	# equivalent FF-pairs (comb. / seq.)	# equivalent PO-pairs (comb. / seq.)	Sim. Time (sec)	Verify Time (sec)
s208	66 / 42	8 / 8	8 (8 / 0)	1 (1 / 0)	1	1
s298	75 / 65	14 / 14	14(14/ 0)	6 (6 / 0)	1	2
s344	114 / 102	15 / 15	15 (15 / 0)	11(11 / 0)	1	1
s349	117 / 101	15 / 15	15 (15 / 0)	11 (11 / 0)	1	1
s382	101 / 92	21 / 21	21 (21 / 0)	6 (6 / 0)	1	1
s386	118 / 60	6 / 6	6 (6 / 0)	7 (7 / 0)	1	1
s400	108 / 88	21 / 21	21 (21 / 0)	6 (6 / 0)	1	1
s420	140 / 84	16 / 16	16 (16 / 0)	1 (1 / 0)	1	1
s444	121 / 85	21 / 21	21 (21 / 0)	6 (6 / 0)	1	1
s510	179 / 115	6 / 6	6 (6 / 0)	7 (7 / 0)	1	1
s526	141 / 106	21 / 21	21 (21 / 0)	6 (6 / 0)	1	1
*s641	128 / 100	19 / 17	17 (2 / 15)	23 (16 / 7)	1	11
*s713	154 / 99	19 / 17	17 (2 / 15)	23 (14 / 9)	1	12
s820	256 / 137	5 / 5	5 (5 / 0)	19 (19 / 0)	1	1
s832	262 / 130	5 / 5	5 (5 / 0)	19 (19 / 0)	1	1
s838	288 / 162	32 / 32	32 (32 / 0)	1 (1 / 0)	1	1
s1196	389 / 273	18 / 18	17 (17 / 0)	14 (14 / 0)	2	3
s1238	429 / 286	18 / 18	18 (18 / 0)	14 (13 / 1)	2	4
s1423	491 / 390	74 / 74	72 (74 / 0)	5 (5 / 0)	16	3
s1488	550 / 309	6 / 6	6 (6 / 0)	19 (19 / 0)	2	3
s1494	558 / 305	6 / 6	6 (6 / 0)	19 (19 / 0)	2	3
*s5378	1074 / 858	164 / 162	162 (142 / 20)	49 (49 / 0)	14	12
s9234	1081 / 599	135 / 135	135 (135 / 0)	39 (39 / 0)	7	5
*s13207	2480 / 1175	490 / 453	419 (379 / 40)	121 (77 / 44)	284	122
*s15850	3379 / 2435	563 / 540	526 (523 / 3)	87 (73 / 14)	545	634
s35932	12492 / 7050	1728 / 1728	1728 (1728 / 0)	320 (320 / 0)	386	75
s38417	8623 / 7964	1464 / 1464	1464 (1464 / 0)	106 (106 / 0)	284	241

* Circuits whose flip-flops counts are reduced after the optimization process.

alent and sequentially equivalent, respectively. Identification of those sequentially equivalent pairs plays an important role in reducing the verification complexity. (4) *# equivalent FF-pairs (comb. / seq.)*: the numbers of equivalent flip-flop pairs that are verified as combinationally equivalent and sequentially equivalent, respectively. Among the 490 / 453 flip-flops of original and optimized s13207, 419 pairs are identified as equivalent using our program, where 379 pairs are combinationally equivalent and 40 pairs are sequentially equivalent. (5) *# equivalent PO-pairs (comb. / seq.)*: the numbers of combinationally and sequentially equivalent primary output pairs. CPU times in seconds are given in the last two columns.

Table 5.2 shows the results of using AQUILA to verify the correctness of ISCAS89 benchmark circuits that were *sequentially* optimized by SIS. Each optimized circuit is obtained by running the following two commands in SIS: (1) Extract the sequential don't care information by FSM traversal *seq_dc_extract*. (2) Optimize the combinational portion of the circuit with sequential don't care information using optimization script *script.rugged*. The circuits in this table are smaller as these two optimization commands cannot handle large designs without partitioning. For most of these circuits, the optimized versions are not combinationally equivalent to their original versions (indicated by asterisk). However, the overall sequential behavior remains the same for each of them. The CPU times of using the verification program in SIS based on the symbolic techniques are listed for comparison.

Table 5.3 shows the AQUILA results for verifying the circuits after sequential redundancy removal. Among the total of 23 circuits, 13 (including s1423, s5378 and s9234) are sequentially equivalent but not combinationally equivalent to their original versions.

5.5 Summary

We have discussed a sequential verifier AQUILA that combines the advantages of ATPG and local BDDs. We show that the assume-and-then-verify concept and the backward justification techniques together provide a very efficient way to explore the similarity between two sequential circuits. When two designs are combinationally equivalent, the computational complexity of this method reduces to combinational verification. On the other hand, if the two designs are structurally different, then this algorithm degenerates to a method that completely relies on backward traversal. To apply the local BDD-based techniques to larger designs, we further use a heuristic called partial justification. Experimental results indicate that this approach has great potential to handle much larger designs than the existing methods.

Table 5.2 Results of verifying ISCAS-89 circuits optimized by SIS script.rugge with sequential don't care information.

Circuit	# nodes orrg / opt	# FFs orig / opt	#internal equiv. pairs (comb / seq)	# equivalent FF-pairs (comb / seq.)	# equivalent PO-pairs (comb / seq)	AQUILA Time (sec)	SIS Time (sec)
s208	66 / 16	8 / 8	15 / 0	8 / 0	1 / 0	1	6
*s298	75 / 22	14 / 14	8 / 10	14 / 0	6 / 0	1	4
*s344	114 / 51	15 / 15	44 / 2	15 / 0	9 / 2	1	10
*s349	117 / 52	15 / 15	41 / 2	15 / 0	9 / 2	1	10
*s382	101 / 40	21 / 21	24 / 7	21 / 0	6 / 0	2	93
*s386	118 / 25	6 / 6	4 / 5	6 / 0	2 / 5	1	2
*s400	108 / 38	21 / 21	22 / 8	21 / 0	6 / 0	2	95
s420	140 / 37	16 / 16	29 / 0	16 / 0	1 / 0	41	3114
*s444	121 / 36	21 / 21	22 / 6	21 / 0	6 / 0	1	62
*s510	179 / 31	6 / 6	21 / 2	6 / 0	6 / 1	5	3
*s526	141 / 40	21 / 21	7 / 23	21 / 0	6 / 0	18	65
*s641	128 / 58	19 / 17	64 / 1	17 / 0	22 / 1	1	19
*s713	154 / 57	19 / 17	70 / 1	17 / 0	22 / 1	1	19
*s820	256 / 56	5 / 5	25 / 7	5 / 0	15 / 4	3	6
*s832	262 / 54	5 / 5	29 / 8	5 / 0	14 / 5	4	6
*s1196	389 / 102	18 / 18	73 / 0	18 / 0	14 / 0	3	17
*s1238	429 / 108	18 / 18	77 / 0	18 / 0	14 / 0	3	17
*s1488	550 / 104	6 / 6	44 / 4	6 / 0	10 / 9	4	10
*s1494	558 / 107	6 / 6	45 / 5	6 / 0	11 / 8	4	10

* Indicates those optimized circuits that are sequentially equivalent, but not combinationally equivalent, to their original versions.

Table 5.3 Results of verifying circuits after sequential redundancy removal.

Circuit	# nodes (orig/opt)	# FFs (orig/opt)	# equiv. internal (comb/seq)	# equivalent FF-pairs (comb/seq)	# equivalent PO-pairs (comb/seq)	Sim. Time (sec)	Verify Time (sec)
s208	66 / 66	8 / 8	65 / 0	8 (8 / 0)	1 (1 / 0)	2	1
*s298	75 / 74	14 / 14	63 / 0	14(11 / 3)	6 (6 / 0)	3	1
*s344	114 / 113	15 / 15	102 / 0	15 (15 / 0)	11(10 / 1)	4	1
*s349	117 / 113	15 / 15	102 / 0	15 (15 / 0)	11 (11 / 1)	4	1
s382	101 / 100	21 / 21	56 / 0	21 (9 / 12)	6 (6 / 0)	4	2
*s386	118 / 111	6 / 6	78 / 0	6 (4 / 2)	7 (2 / 5)	4	1
*s400	108 / 101	21 / 21	55 / 0	21 (9 / 12)	6 (6 / 0)	4	2
s420	140 / 140	16 / 16	139 / 0	16 (16 / 0)	1 (1 / 0)	5	1
*s444	121 / 105	21 / 21	49 / 0	20 (9 / 11)	6 (6 / 0)	4	7
s510	179 / 179	6 / 6	172 / 0	6 (6 / 0)	7 (7 / 0)	6	1
*s526	141 / 139	21 / 21	90 / 0	20 (11 / 9)	6 (6 / 0)	6	20
*s641	128 / 120	19 / 18	97 / 0	17 (2 / 15)	23 (16 / 7)	6	2
s713	154 / 111	19 / 19	79 / 0	19 (19 / 0)	23 (23 / 0)	6	2
s820	256 / 256	5 / 5	237 / 0	5 (5 / 0)	19 (19 / 0)	10	1
s832	262 / 256	5 / 5	217 / 0	5 (5 / 0)	19 (19 / 0)	10	1
s838	288 / 288	32 / 32	287 / 0	32 (32 / 0)	1 (1 / 0)	11	1
*s1196	389 / 387	18 / 18	364 / 0	17 (17 / 0)	14 (12 / 2)	13	2
*s1238	429 / 386	18 / 18	305 / 0	17 (17 / 0)	14 (12 / 2)	14	8
s1488	550 / 550	6 / 6	550 / 0	6 (6 / 0)	19 (19 / 0)	19	3
s1494	558 / 548	6 / 6	523 / 0	6 (6 / 0)	19 (19 / 0)	17	2
*s1423	491 / 463	74 / 74	441 / 0	71 (68 / 3)	5 (5 / 0)	11	96
*s5378	1074 / 919	164 / 139	891 / 18	139(139 / 0)	49 (43 / 6)	57	120
*s9234	1081 / 1067	135 / 131	1052 / 0	125 (122 / 3)	39 (38 / 1)	140	46

* Indicates those redundancy-removed circuits that are sequentially equivalent, but not combinationally equivalent, to their original versions.

Chapter 6

Algorithm for Verifying Retimed Circuits

In this chapter, we discuss a heuristic for verifying retimed circuits. Usually after the retiming transformation, some flip-flops or latches in a sequential circuit are re-positioned. Thus, the behavior of the circuit's combinational portion is changed. However, it has been proven that the overall sequential behavior of a retimed circuit is equivalent to its original version with respect to the notion of 3-valued equivalence [94,119]. Based on the framework described in the previous two chapters, a heuristic called "*delay compensation*" is developed to further improve the computational efficiency of verifying retimed circuits.

6.1 Introduction

The retiming technique [90,117] moves latches forward or backward to reduce the clock cycle time or to minimize the number of latches. Because retiming can be preformed on a circuit without a known reset state, the theoretical correctness of retiming is not obvious. The theoretical correctness of retiming for a sequential circuit without a reset state has recently been re-examined more thoroughly [94,119].

A retiming transformation can be broken down into a sequence of atomic retiming moves. There are four types of atomic moves as shown in Fig. 6.1. Among them, the stem-to-branch move (type 3) that moves a latch from a

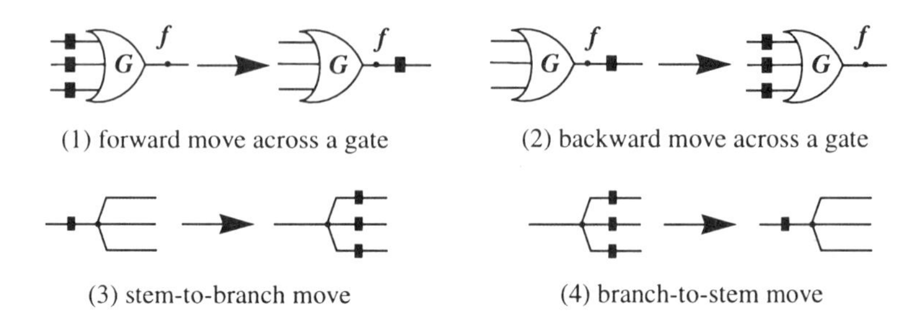

Fig. 6.1 Atomic retiming moves.

multiple-fanout stem to its branches may violate some stringent definitions of sequential equivalence such as safe replaceability [105]. However, it has been proven that a retimed circuit is functionally equivalent to its original design *after a certain number of arbitrary input vectors is applied.* According to this definition, the output responses during the application of these vectors are ignored. Also, it was proven that retiming moves including the stem-to-branch moves are safe transformation with respect to the definition of 3-valued equivalence [94,119]. The later property is particularly important because most designers rely on the 3-valued logic simulation to verify their designs.

We discuss a pre-processing algorithm to generalize the verification framework for retimed circuits. The overview of the verification process for a retimed circuit is shown in Fig. 6.2. Before applying the incremental verifier, the miter is first pre-processed and converted into an intermediate form. In this pre-processing, more equivalent signal pairs are created through a technique called *delay compensation* to improve the computational efficiency of the subsequent incremental verification process.

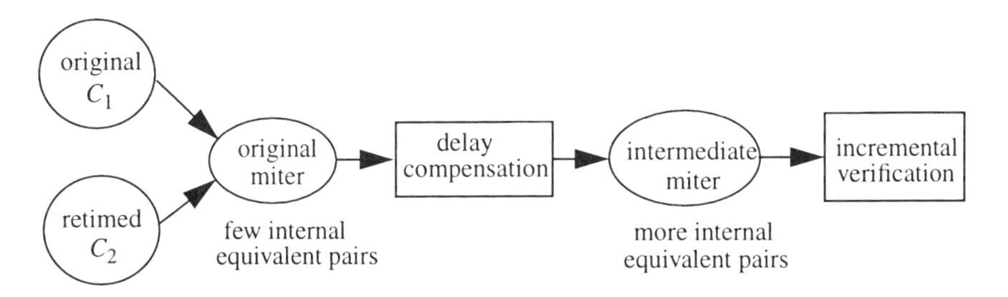

Fig. 6.2 Flow of the verification process for retimed circuits.

6.2 Pre-Processing Algorithm

The computational efficiency of incremental verification is strongly related to the degree of structural similarity between the two circuits under verification. In general, very few equivalent signal pairs exist between the retimed circuit and its original version because the combinational portion of the circuit is dramatically changed by retiming. Therefore, the incremental verification may perform poorly for retimed circuits. However, it is observed that there may exist many *delayed-equivalent signal pairs* (defined below) between the two circuits. Identification and proper utilization of these delayed signal pairs can significantly speed up the process.

Definition 6.1 (*delayed-equivalent pair*) A signal pair (f_1, f_2) is called a delayed-equivalent pair if the responses of f_1 are equivalent to the delayed version of the responses of f_2 (by a number of clock cycles D). Signal f_1 and f_2 are called the *delayed signal* and the *ahead signal* of the pair.

Consider the example in Fig. 6.3 [117]. We use the notation of a and a' to represent the corresponding signals in the original and the retimed circuits, respectively. In this example, signals d' and e' after retiming remain equivalent to their corresponding signals d and e in the original circuit. But the signal pair (a, a') becomes delayed-equivalent because a primary input, say x_1, reaches a and a' in different numbers of clock cycles.

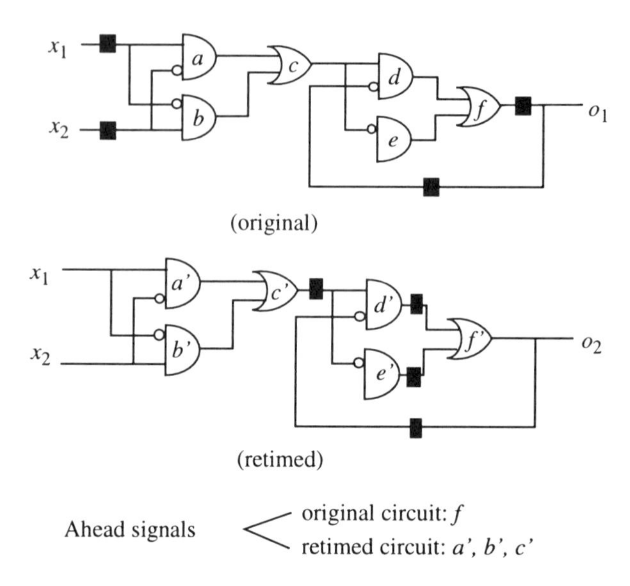

Fig. 6.3 An example illustrating delayed equivalent pairs and ahead signals.

Signal pairs (b, b'), (c, c') and (f', f) are delayed-equivalent as well. We attempt to convert these delayed-equivalent signal pairs back into equivalent pairs by moving the latches through retiming moves in the pre-processing stage. The pre-processed circuits are then verified for equivalence using the procedure described in the previous chapters. Note that the pre-processing to be described is not a reverse-retiming of the retimed circuit because of three reasons: (1) The retiming moves are independent of the retiming program producing the retimed circuits. (2) The compensation moves are applied to both circuits C_1 and C_2 (not just to the retimed circuit). (3) We only move the latches backward. This type of moves has been proven to be valid even with respect to more stringent definitions of circuit equivalence such as safe replaceability [105].

The pre-processing consists of three main steps:

(1) For each signal f and every primary input x_i, compute the minimal number of latches from x_i to f for all paths between them. This

number, an integer denoted as $\sigma(x_i, f)$, is referred to as the *signature* of the signal f with respect to the primary input x_i.

(2) Identify candidate delayed-equivalent pairs based on the signatures and the simulation results.

(3) *Compensate* the ahead signals of the candidate delayed-equivalent pairs by moving latches at the outputs of the ahead signals backward.

The details of these three steps are described next.

6.2.1 Signature Computation

The signature of each internal signal with respect to a given primary input can be computed in linear time. This is done through a process similar to the wavefront propagation that starts from the given primary input and propagates forward towards primary outputs. The propagation stops when every signal that is reachable from the given primary input has been visited. Each signal is associated with a cycle number. Initially the cycle number of the given primary input is set to 0 and all others are set to infinity (indicating the status of being not visited yet). At any time, the signal with the smallest cycle number in the frontier is selected to further advance the wavefront. When the wavefront propagates across a latch, the cycle number is incremented by one, otherwise, the cycle time is assigned to the newly visited frontier signal. After this wavefront propagation, a signal's signature is simply the associated cycle number. Since no signal is visited more than once, this algorithm has linear-time complexity in terms of the circuit size.

6.2.2 Deriving Candidate Delayed-Equivalent Pairs

After signature computation, candidate equivalent or delayed-equivalent pairs between the original circuit and the retimed circuit can be derived using functional simulation, name comparison, and/or user-provided information. A signal pair (f_1, f_2) is considered a candidate delayed-equivalent pair if two criteria are satisfied: (1) (f_1, f_2) is consistent (Definition 6.2). (2)

The output responses of one of them is equivalent to the other but delayed by D cycles, where D is the difference between the signatures of f_1 and f_2.

Definition 6.2 (*consistent pair*) A signal pair (f_1, f_2) is called a consistent signal pair if for every primary input x_i, signature $\sigma(x_i, f_1)$ and $\sigma(x_i, f_2)$ are both infinity (i.e., not reachable from x_i) or $\sigma(x_i, f_1) - \sigma(x_i, f_2) = D$, where D is an integer constant (i.e., the signature differences with respect to all primary inputs are identical). The ahead signal in this pair (i.e., f_1 if D is negative and f_2 if D is positive) is called the *candidate ahead signal* with delay size D.

Inconsistent pairs are not considered as candidate pairs because when an atomic forward (or backward) retiming move across a gate is performed, the signature of a gate's output with respect to *every* reachable primary input is incremented (or decremented) by 1 at the same time. Hence, a signal pair that has different signature differences with respect to different primary inputs cannot be a delayed-equivalent pair.

Example 6.1 Consider a circuit with 3 primary inputs and a signal pair (f_1, f_2), where f_1 and f_2 are signals in the original and the retimed circuits, respectively. Let the signature vector (signatures with respect to the three primary inputs) for f_1 and f_2 be $\sigma(f_1) = (0, 2, 3)$ and $\sigma(f_2) = (0, 1, 2)$. The signature difference of this pair with respect to the first primary input is $0 - 0 = 0$, while those of the second and the third primary inputs are 1. Therefore, this pair is not a consistent pair, and it will be removed from the list of candidate delayed-equivalent pairs.

Example 6.2 Suppose (f_1, f_2) is a signal pair, and the signature difference between f_1 and f_2 is 1 for every primary input (i.e., $\sigma(x_i, f_1) - \sigma(x_i, f_2) = 1$). If the logic values of f_1 and f_2 in response to the functional input sequence of 6 vectors using 3-valued logic simulation are $(u, \underline{u, 0, 1, 1, 0})$ and $(\underline{u, 0, 1, 1, 0}, 0)$, respectively, then (f_1, f_2) is considered as a candidate delayed-equivalent pair because the response of f_1 is equivalent to that of f_2 but is delayed by one clock cycle (the identical part of the responses are underlined), which is identical to their signature difference.

6.2.3 Delay Compensation

Once the candidate equivalent and delayed-equivalent pairs are derived, a heuristic is applied in an attempt to convert the delayed-equivalent pairs into equivalent pairs in the miter. This heuristic performs backward retiming moves to systematically reduce the delay size of each candidate delayed-equivalent pair to zero.

In this heuristic, the signals in the combinational portion of the miter are first sorted in a fanout-first order (i.e., every node is placed after its transitive fanout). A simple iterative process is then employed. Each iteration goes through the sorted signal list and processes each signal according to four rules:

(Case 1): The current signal f is the *ahead signal* of a candidate delayed-equivalent pair with delay size D. Assume f connects to k latches (e.g., signal f connects to 2 latches in the left figure of Fig. 6.4(a)). If $k < D$, then we move all k latches backward across this gate. After this retiming move, the delay size of the candidate pair in which the current signal is involved is decremented by k. On the other hand, if $k \geq D$, then only D latches are moved backward across it. The delay size of the current candidate pair is reduced to 0 in this case, and the candidate pair becomes a candidate equivalent pair. An example is shown in Fig. 6.4(a) where f is ahead by 1 clock cycle but fans out to 2 latches.

(Case 2): If the current signal is a delayed signal of a candidate delayed pair, then no action is taken. We do not attempt to make any forward retiming move.

(Case 3): If the current signal is in a candidate equivalent pair, then no action is taken.

(Case 4): If the signal does not belong to any candidate pair (neither delayed equivalent nor equivalent), then all latches (if any) at the output of the signal are moved backward from its fanouts to its fanins. This move attempts to create more latches at the outputs of its fanin signals such that the subsequent processing of its fanin signals will have enough latches available for case 1 compensation move.

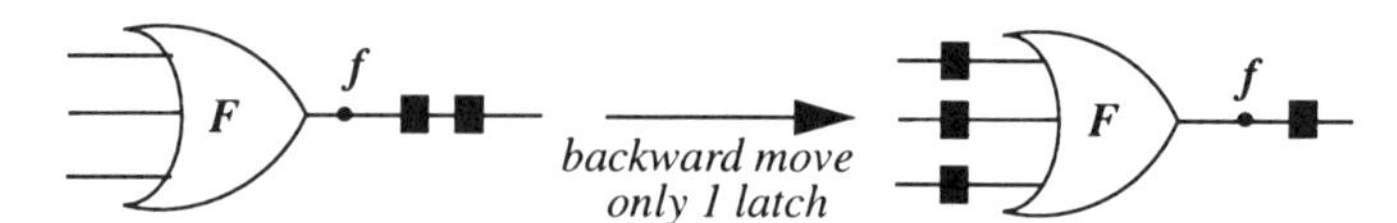

(a) *f* belongs to a candidate delayed pair and is **ahead** by 1 cycle.

F *f* *no action*

(b) *f* is the delayed signal of a candidate delayed pair.

F *f* *no action*

(c) *f* is in a candidate equivalent pair.

(d) *f* does not belong to any candidate delayed pair.

Fig. 6.4 Rules for compensation move.

This process iterates until no latches can be moved backward. Fig. 6.5 shows an outline.

6.3 Experimental Results

We applied this heuristic to verify the retimed circuits generated by the retiming program in SIS. For each ISCAS-89 benchmark circuit, we first performed the logic optimization using SIS script *script.rugged*. We then used the command "*retime -n -i*" to retime the circuit without assuming a reset state. The program consisting of the pre-processor and the incremental verifier was then used to verify the optimized-and-retimed circuit against its original version. Table 6.1 summarizes the results. The numbers of the connections and the latches of the original and the retimed circuits are

```
Delay-compensation-for-retiming-verification(C1, C2)
{
    1. Construct the joint network of C1 and C2 by connecting PIs together.
    2. Foreach primary input, compute the signature of each signal.
    3. Derive candidate pairs (by functional simulation and name comparison).
    4. Compute the delay size of each consistent delayed pair.
    5. Sort the signals in a fanout-first order.
          (based on the topology of the miter's combinational portion)
    6. While(1){
          foreach signal f in the joint network{
                do compensation move on f if possible;
                update the delay size of a pair if necessary;
          }
          if (no latch could be moved backward) break;
    }
    7. Return the transformed miter;
}
```

Fig. 6.5 The flow of delay compensation.

shown in the second and third columns, respectively. The CPU times for a Sun Sparc-5 workstation equipped with 128 M byte memory are shown in the last column. These CPU times include the pre-processing part. Because the circuits are *optimized* and then retimed, the pre-processing heuristic will not be able to convert the two circuits into two combinationally equivalent ones. However, a careful examination of the results shows that this heuristic indeed creates many new equivalent pairs that did not exist in the original circuits, and thereby reduces the computational complexity of the subsequently incremental verification substantially. It is worth mentioning that this method does not rely on any information from the retiming program of SIS.

As mentioned in Chapter 4, the modified ATPG for verification adopts a breadth-first strategy, instead of a depth-first search strategy. This modification makes it more efficient for verifying equivalence than a traditional sequential ATPG, but less suitable for finding a distinguishing sequence when one exists. As a result, it is beneficial to integrate our program with a sequential ATPG program as a dual-program to handle different situations.

For example, our verification program combining the local BDD and ATPG techniques can verify the correctness of the optimized-and-retimed circuit s344 in 1 second, while STG3 (a sequential ATPG) [35] cannot reach the conclusion that the miter's outputs are all undetectable (which is only the necessary condition) after searching for 10 hours. On the other hand, STG3 found a 3-valued distinguishing sequence with two input vectors for s5378

Table 6.1 Results of verifying circuits optimized-and-then-retimed in SIS.

Circuit	# connections original / optimized+retimed	# FFs original / optimized+retimed	3-valued equivalent (?)	CPU-Time (seconds)
s27	9 / 9	3 / 3	YES	1
s208*	66 / 16	8 / 9	YES	1
s298*	81 / 29	14 / 23	YES	3
s344*	114 / 55	15 / 27	YES	1
s349*	117 / 56	15 / 27	YES	1
s382*	105 / 45	21 / 30	YES	2
s386*	118 / 29	6 / 14	YES	2
s400*	112 / 43	21 / 23	YES	2
s420*	140 / 33	16 / 19	YES	65
s444*	125 / 42	21 / 25	YES	2
s510	179 / 31	6 / 6	YES	4
s526*	147 / 49	21 / 39	YES	45
s641	130 / 60	19 / 17	YES	2
s713	156 / 59	19 / 17	YES	2
s820	256 / 56	5 / 5	YES	3
s832	262 / 54	5 / 5	YES	3
s1196	390 / 103	18 / 18	YES	6
s1238	430 / 109	18 / 18	YES	6
s1488	550 / 105	6 / 6	YES	6
s1494	558 / 108	6 / 6	YES	6
s1423*	491 / 141	74 / 86	YES	8
s5378*	1074 / 392	164 / 343	NO	100.6 (STG3)
s9234.1	1081 / 229	211 / 211	YES	29

* Circuits whose clock cycle times have been improved by the retiming-program in SIS.

and its SIS-retimed circuit in less than two minutes. For this case, our verification program takes longer time to find a distinguishing sequence because of the breadth-first search strategy. For retimed circuit s5378, we are not sure whether or not the error comes from SIS. There is a possibility that it was introduced when we translated the circuit format.

6.4 Summary

We have discussed a delay-compensation heuristic to reduce the computational complexity of verifying retimed circuits. This heuristic is based on an idea of converting the delayed-equivalent signal pairs into equivalent signal pairs through a guided process that moves the latches backward. The experimental results provide evidence that this system can be used to efficiently verify the equivalence of circuits after intensive optimization and retiming.

Chapter 7

RTL-to-Gate Verification

In this chapter we discuss the use of a gate-to-gate equivalence checker to verify a gate-level implementation against its structural RTL specification. We begin with the basic methodology for RTL-to-gate verification and then discuss two key issues in this application: (1) external don't care modeling, and (2) integration of the symbolic FSM traversal with the incremental verification.

7.1 Introduction

RTL-to-gate verification can be performed in two steps. First, a fast-path synthesis (i.e., synthesis/optimization with a low effort) of the RTL specification is performed to derive a gate-level specification. Then, a gate-to-gate equivalence checker is called to compare the gate-level specification with the gate-level implementation. One serious problem with this approach is the *false negative* problem – the verifier may report that these two gate-level circuits are inequivalent, even though the gate-level implementation correctly realizes the specified behavior of the RTL specification. This false negative problem arises from the undefined behavior associated with an RTL specification. We refer to an input sequence as an *external don't care* (sequence) if the output response to this sequence is not defined in the RTL specification. Otherwise, it is called a *care* input sequence. External don't cares could be used for design optimization [11]. Different interpretations of these don't cares result in different logic implementations. To resolve the false negative problem, the don't cares in the RTL specification need to be extracted. These extracted don't care conditions are then characterized as a sequential single-output network. This network is then incorporated into the miter model for a gate-to-gate verification to avoid the false negative problem. The details will be described in Section 7.2.

Incremental verification is particularly efficient for verifying two sequential circuits with similar state encodings. For circuits with different state encodings, e.g., a minimal-bit encoded circuit and a one-hot encoded circuit, this approach is not necessarily advantageous. A hybrid technique combining the incremental verification with the symbolic finite state machine traversal is introduced to circumvent this limitation. This method divides the joint network of the specification and the implementation (the miter) into two portions: (1) one portion in which the specification and the implementation have similar state encodings, and (2) another portion in which the state encodings differ. Then, the incremental technique and the FSM traversal technique are integrated in such a way that the former is applied to portion (1) and the latter is applied to portion (2). For instance, consider a processor-like design with a data-path and a controller. Throughout design optimization, the data-path portion, containing arithmetic units, counters, shift-registers, etc., is rarely re-encoded, while the state encoding of the controller might be changed. In the hybrid method, the symbolic FSM traversal is used for the differently encoded portion to derive a subset of unreachable states. These unreachable states are then intelligently used to speed up the incremental verification process that follows. By integrating the FSM traversal and the incremental verification together, the hybrid approach is more likely to verify designs that are beyond the capability of a simple approach.

7.2 Don't Care Modeling

An RTL specification is most likely to be incompletely specified. Incomplete specification implies external don't care conditions. Two gate-level implementations, both conforming to an RTL specification, may have different Boolean behaviors due to the different interpretations of the external don't care conditions. Therefore, directly using a gate-to-gate equivalence checker to verify whether a gate-level implementation conforms to its RTL specification may result in a false negative result. To resolve this problem, one can generalize the miter model for verification as shown in Fig. 7.1. In this generalized model, the external don't care set

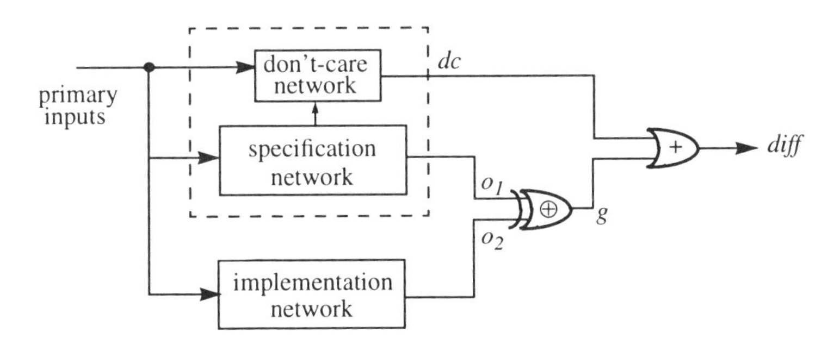

Fig. 7.1 Generalized miter considering external don't care set.

extracted from the RTL specification is represented as a sequential network with only one primary output *dc*. This don't care network is considered as a part of the specification and could share the flip-flops with the specification network. The specification network here is a gate-level network produced from the RTL specification through some low-effort synthesis. If an input sequence is a don't care sequence, then the don't care network produces a value '1' at signal *dc*. A value '0' indicates the applied input sequence is a care sequence. For example, an 8-bit add-and-shift multiplier produces one care output for every 8 clock cycles. Therefore, the don't care network should produce an output sequence, 11111110, which indicates that only at the eighth cycle, the output is a care output. In this generalized miter, every output pair of the specification and the implementation is still connected to an exclusive-OR gate (denoted as g). This signal is further connected to a 2-input OR gate with signal dc as the other input. The output of this OR gate is denoted as signal *diff*.

Property 7.1 Assume both the specification and the implementation networks have a known reset state. Signal g stuck-at-0 fault in Fig. 7.1 is untestable if and only if signal o_1 is equivalent to o_2 for all care input sequences.

An input sequence D must satisfy two conditions to be a valid distinguishing sequence for (o_1, o_2): (1) D should differentiate (o_1, o_2), i.e., set signal g to '1'; and (2) D should be a care sequence, i.e., set signal dc to '0'. These two conditions are exactly the same as the requirements of a test for the signal g stuck-at-0 fault. Therefore, this necessary and sufficient condition can be directly examined by a sequential ATPG program to prove the equivalence or to find a distinguishing sequence for a target signal pair. Furthermore, the idea of incremental verification discussed in Chapters 4 and 5 are still applicable to this generalized model.

In the above discussion, we assume that both networks have a known reset state. It can be generalized for the case when the two networks do not have a reset state.

Property 7.2 Consider the model in Fig. 7.1. Assume that the specification and the implementation networks do not have a known reset state. Signal o_2 is a *3-valued safe replacement* of signal o_1 if there does not exist an input sequence D and a state s_2 in the implementation network such that the responses at signal dc, o_1, and o_2, derived using the conservative 3-valued logic simulation, simultaneously satisfy two conditions: (1) $dc(x, D) = 1$, and (2) $(o_1(x, D), o_2(s_2, D))$ belongs to $\{(0, 1), (1, 0)\}$, where x denotes the unknown initial state (i.e., all flip-flops have an initial value u).

The don't cares can be classified as either *external don't cares*, or *internal don't cares*. The external don't cares are due to the undefined behavior, while the internal don't cares are the *impossible value combinations at internal signals*. For example, the unreachable states are internal don't cares. The internal don't cares are due to the temporal and/or spatial correlations among signals in the design. In the context of design optimization, the external and internal don't cares need not be differentiated. But, for verification, they are different in essence. Unlike external don't cares, not considering internal don't cares will not cause a false negative problem. However, extracting the internal don't cares to bound the search during verification could substantially reduce the computational complexity. For example, whenever the backward search performed in ATPG reaches an unreachable state, it can backtrack immediately to avoid unnecessary search. The external and some internal don't cares can be extracted from the RTL specification and represented in a joint don't care network. Therefore, the inputs of this network, in general, include both primary inputs and some internal signals of the specification network. The don't care network can also share some internal signals with the specification.

7.3 Integration with FSM Traversal

Incremental verification and symbolic FSM-traversal based verification have different strengths and weaknesses. Incremental verification is effective for verifying designs with a great deal of internal similarity. Symbolic FSM-traversal is structure-independent but is vulnerable to the memory explosion problem. It is likely that neither approach will work for verifying a large gate-level network synthesized from an RTL specification against a customized network with different state encoding for part of the design. For a processor-like design in which the specification and the implementation have differently encoded controllers, even though significant similarity exists in the data-path, the incremental verification approach may not be effective due to the dissimilarity of the controllers. On the other hand, the FSM-traversal approaches will run into the memory explosion problem due to the huge state space, mainly contributed by the data-path registers. In the following, we introduce a hybrid method that combines the advantages of these two approaches for dealing with large designs with re-encoded controllers.

7.3.1 Computing a Subset of Unreachable States

In this subsection, we describe a procedure using FSM traversal on the *differently encoded portion of the miter* to compute a subset of unreachable states. In the following, we first describe how to automatically identify the differently-encoded portion in the respective network. Given a specification with a don't care network and an implementation, we first simulate a large number of random input sequences or design verification vectors, if available, to pair up the candidate equivalent flip-flops. A flip-flop is paired up with another flip-flop in the other network if the two flip-flops have identical responses to all applied care input sequences. The paired flip-flops are classified as *state-inessential* FFs. On the other hand, if a flip-flop is not paired with any other flip-flop, it is classified as a *state-essential* flip-flop. Fig. 7.2 shows an example. The specification and don't care network has

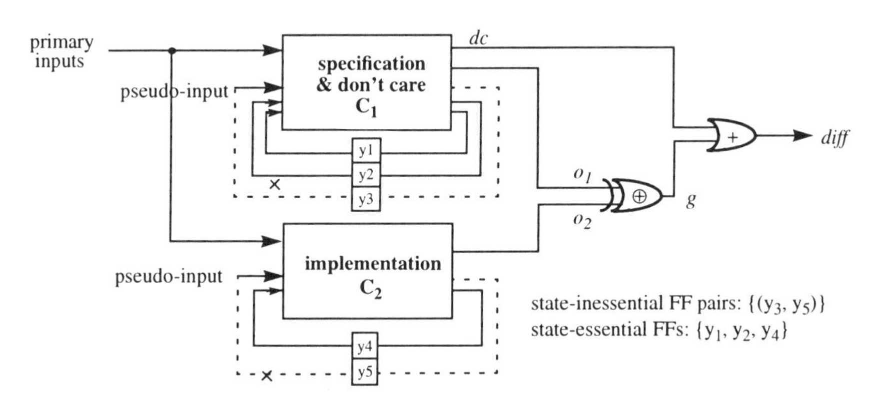

Fig. 7.2 An abstract miter for computing partial set of unreachable states.

three flip-flops $\{y_1, y_2, y_3\}$, and the implementation has two flip-flops $\{y_4, y_5\}$. Suppose (y_3, y_5) is a candidate flip-flop pair after simulation. The set of state-inessential FFs will be $\{y_3, y_5\}$, and the set of state-essential FFs will be $\{y_1, y_2, y_4\}$. Based on this classification, we construct an *abstract miter* for performing the FSM traversal. This abstraction removes the state-inessential flip-flops and treats their outputs as pseudo-inputs to the miter (illustrated in Fig. 7.2). A FSM traversal is then performed to derive a set of unreachable states (specified over $\{y_1, y_2, y_4\}$). These unreachable states are then used as internal don't cares to bound the search in the subsequent incremental verification process.

We explain the process of computing a subset of unreachable states from the abstract miter using the example in Fig. 7.2. For this miter, the primary input set is $\{x\}$, the pseudo primary input set is $\{y_3, y_5\}$, and the set of present state lines is $\{y_1, y_2, y_4\}$. We denote the set of next state lines of the state-essential FFs as $\{Y_1, Y_2, Y_4\}$, and their functions in terms of the present state lines, primary inputs, and pseudo primary inputs as $\{f_1, f_2, f_4\}$. Let the Boolean function for signal *dc* be F_{dc}. The transition relation for this abstract miter considering external don't care set, F_{dc}, can be expressed as a Boolean function:

$$Transition\text{-}relation = (\exists x y_3 y_5)((Y_1 \equiv f_1) \cdot (Y_2 \equiv f_2) \cdot (Y_4 \equiv f_4) \cdot (\overline{F_{dc}}))$$

In this expression, all state-inessential FFs (i.e., y_3 and y_5) are existentially quantified in addition to the primary input variables. Also, the transition relation is intersected with the *care function* $\overline{F_{dc}}$ to guarantee that the transition occurs in the care input space. Based on this transition relation, the next states of a given set of states can be computed in one-shot through image computation as described in Chapter 2. A breadth-first traversal of the state transition graph from the reset state is employed to compute the reachable states of this abstract miter. Finally a negation on this set is computed to derive the unreachable states. It is guaranteed that the unreachable states derived in this procedure are also unreachable states for the original miter. For instance, if $(y_1, y_2, y_4) = (1, 1, 1)$ is an unreachable state in the abstract miter, then $(y_1, y_2, \mathbf{y_3}, y_4, \mathbf{y_5}) = (1, 1, -, 1, -)$ is an unreachable set of states in the original miter.

7.3.2 Incremental Verification with Don't Cares

Consider an internal signal pair (a_1, a_2) under equivalence checking. The don't care network poses a restriction on the input space when deriving a distinguishing sequence for this pair. We have earlier shown that a sequential ATPG program can be directly used to check the equivalence of this pair under such a restriction (Fig. 7.3). In the following, we discuss the

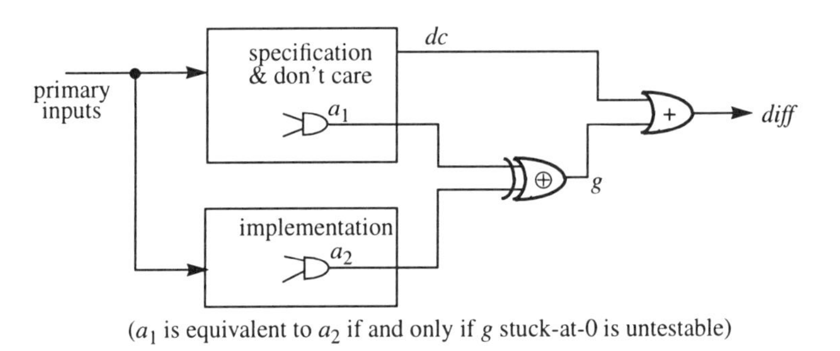

Fig. 7.3 A generalized miter for checking equivalence of (a_1, a_2).

use of local BDDs to check the equivalence of a signal pair under don't care constraints. As illustrated in Fig. 7.4, a local cutset λ in the fanin cones of signals a_1, a_2, and *dc* is first selected. Then, the BDDs representing signals *dc* and *g* in terms of λ, denoted as $F_{dc(\lambda)}$ and $F_{g(\lambda)}$, are constructed. If $(F_{g(\lambda)} \cdot \overline{F_{dc(\lambda)}})$ is tautology zero, it can be concluded that a_1 and a_1 are equivalent. Otherwise, the cutset can be dynamically expanded backwards toward primary inputs, followed by the BDD construction of $(F_{g(\lambda)} \cdot \overline{F_{dc(\lambda)}})$ with respect to the new cutset. This process continues until either a conclusion is possible or we reach a pre-specified limit on the number of backward expansions.

Suppose F_u denotes the characteristic function of the unreachable states in terms of present state lines derived in the FSM traversal phase. We can

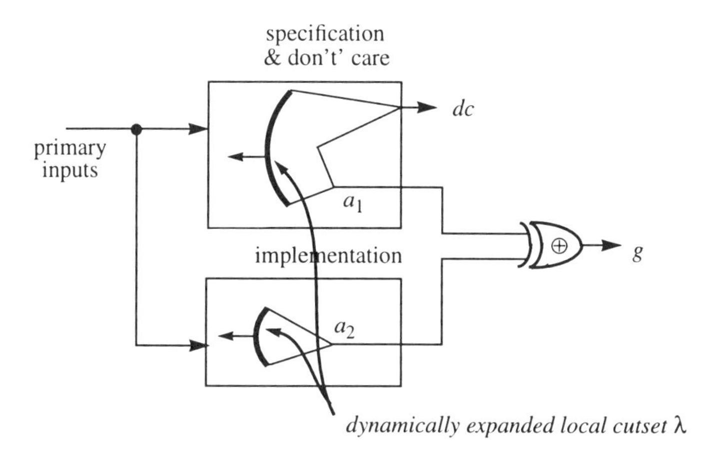

Fig. 7.4 Checking equivalence of (a_1, a_2) using local BDDs (considering don't cares).

further use F_u to speed up the equivalence checking process with respect to a selected cutset λ. First, those present state variables present in F_u but not present in λ are *universally quantified.* Let the function of the quantified unreachable states be F_{qu}, which is a function in terms of λ only. If $(F_{g(\lambda)} \cdot \overline{F_{dc(\lambda)}} \cdot \overline{F_{qu(\lambda)}})$ is tautology zero then the target pair is equivalent. Intuitively, the unreachable states characterized by F_u, representing the set of impossible value combinations at the present state lines, can be used to derive the impossible value combinations at the selected cutset λ. Function F_{qu} characterizes a subset of such impossible value combinations at λ. The vectors in terms of λ distinguishing the target signal pair in the care space are characterized by function $(F_{g(\lambda)} \cdot \overline{F_{dc(\lambda)}})$. Furthermore, function $(F_{g(\lambda)} \cdot \overline{F_{dc(\lambda)}} \cdot \overline{F_{qu(\lambda)}})$ represents the set of these *care distinguishing vectors* that are *reachable* from the reset state in terms of λ. Therefore, if this set is empty, then the target signal pair is equivalent.

7.4 Overall Flow for RTL-to-Gate Verification

Fig. 7.5 summarizes the overall flow. The RTL specification is first synthesized using a low-effort optimization. External don't cares extracted from the specification are represented as a don't care network. The entire procedure uses 3 different miter models: (1) the original miter, (2) the reduced miter, and (3) the abstract miter. The reduced miter is constructed for identifying the equivalent flip-flop pairs as discussed in Chapter 5. It explicitly connects the present state lines of each *candidate* pair together. Based on this reduced miter, an abstract miter is further constructed for the FSM traversal to compute a subset of unreachable states. This step involves treating the inputs (outputs) of every candidate flip-flop as pseudo primary outputs (inputs).

In summary, there are four major steps in this iterative process: (1) Construct the reduced miter. (2) Construct the abstract miter and perform the FSM traversal. (3) Perform incremental verification to check equivalence of each candidate internal signal pair in the reduced miter. (4) Filter out the false candidate flip-flop pairs. If no false candidate flip-flop pair exists, then the iterative process terminates. Finally, the equivalence of each primary output pair is checked by using the local BDD-based and/or the ATPG-based methods.

7.5 Experimental Results

The RTL-to-gate verification techniques have been incorporated into the equivalence checker AQUILA. To evaluate its capability for handling re-encoded circuits with an external don't care set, a 2's complement add-and-shift multiplier [56] is used as a test case. The data-path of the multiplier is described at the micro-architecture level consisting of a number of RTL components. The controller is described by its state transition table with a number of unspecified transitions. The external don't cares associated with the controller are extracted from the incompletely specified functional specification. Some of these don't cares are due to the assumption of the environment, and others are due to the interaction between the data-path

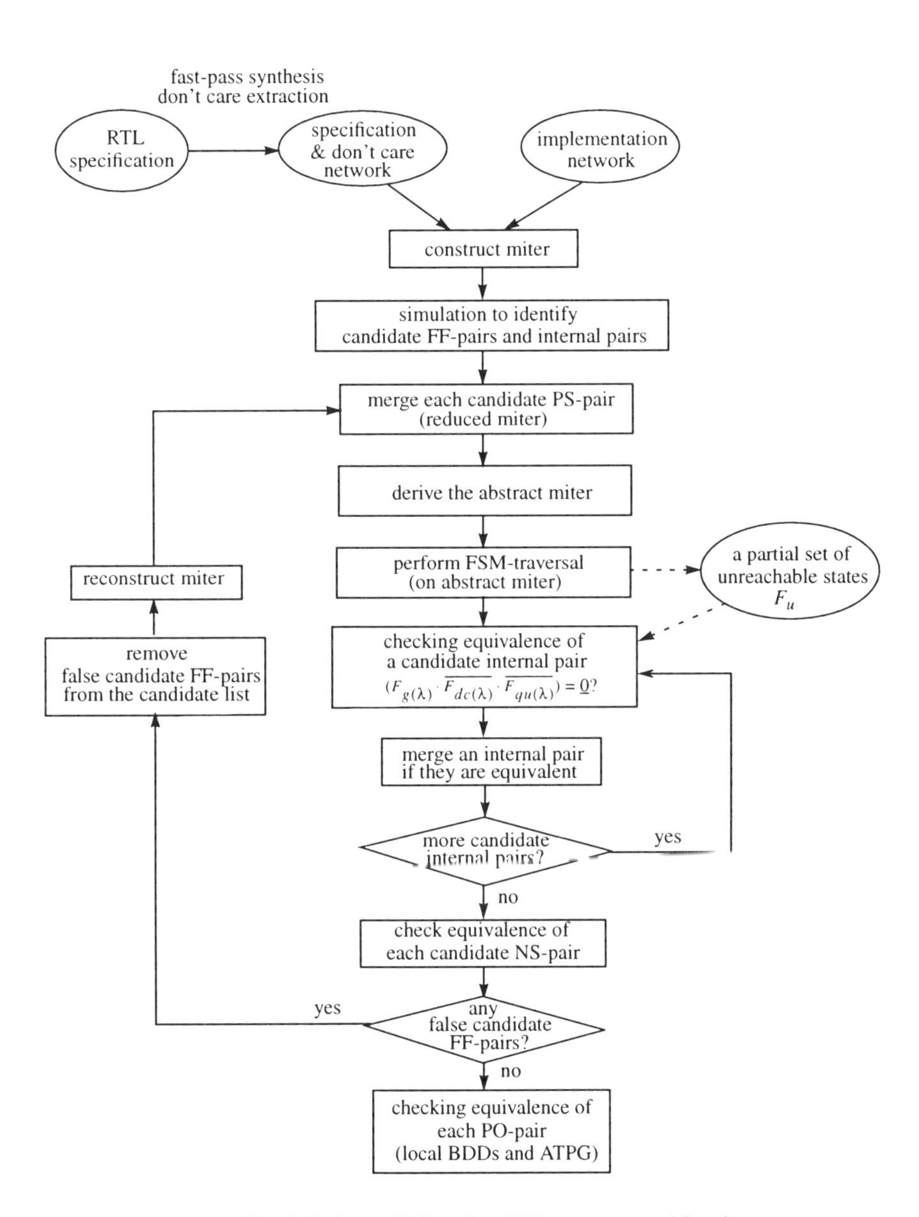

Fig. 7.5 Overall flow for RTL-to-gate verification.

and the controller. The state table of the controller is then encoded by the state encoding program, *nova* [121], to generate the specification network and the don't care network. A different implementation is further created by using the one-hot encoding to synthesize the controller. As a consequence, the latter implementation has 3 more flip-flops than the specification network. If correctly synthesized, the cycle behaviors of these two circuits for all care sequences should be identical. The entire flow of this experiment is illustrated in Fig. 7.6.

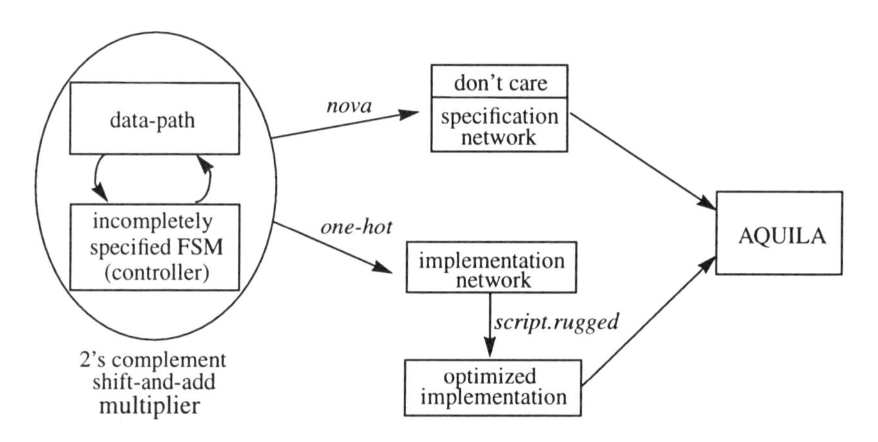

Fig. 7.6 The flow of verifying a 2's complement multiplier against its structural RLT specification.

Because the specification network and the implementation network interpret the external don't care conditions differently, they are reported as inequivalent by simulation when the don't cares are not taken into account. AQUILA successfully proves that the optimized implementation is indeed equivalent to the specification in the care space. The results are summarized in Table 7.1. The first 4 columns show the circuit statistics. The last 2 columns show the verification times using the verification program in SIS (based on symbolic FSM traversal) and AQUILA. For SIS verification, we report the times for verifying two identical implementation circuits because SIS does not deal with the don't care issues. Due to the huge state space

Table 7.1 Results of verifying shift-and-add multipliers with external don't cares.

# Bits	# FFs spec. / impl.	# Nodes spec / impl	# Literals spec. / impl.	SIS (sec)	AQUILA (sec)
4-bit	19 / 22	147 / 119	469 / 293	*6.8	1.8
16-bits	57 / 60	421 / 346	1249 / 846	—	3.4
32-bits	105 / 108	745 / 613	2214 / 1483	—	7.0

*: verification time of identical circuits in SIS.
—: cannot complete because of memory explosion on a 128 Mbyte workstation.

created by the data-path registers, the symbolic FSM traversal runs into the memory explosion problem on a SUN Sparc-5 workstation with 128 Mbyte memory for the 16-bit and 32-bit add-and-shift multipliers. AQUILA completes the verification tasks in several seconds. Increasing the bit-width of the data-path does not cause significant increase in the computational complexity.

Table 7.2 shows the experimental results of verifying a simple micro-processor design, which contains a register file, a decoder, a data-path, and a controller. The original specification is described in Verilog at the RT-level. We re-encoded the controller using the same number of state bits for a new implementation. We then synthesized and optimized both the specifi-

Table 7.2 Results of verifying a re-encoded simple processor-like designs.

Data-width	#FFs spec. / impl.	#Nodes spec / impl	#Literals spec. / impl.	SIS (sec)	AQUILA (sec)
8-bit	94 / 92	1977 / 536	4363 / 1240	1887	7.7
16-bits	174 / 174	3761 / 979	8027 / 2271	—	15.7
21-bits	224 / 224	4877 / 1259	10317 / 2916	—	38.3

—: cannot complete because of memory explosion on a 128 Mbyte workstation.

cation and the implementation using SIS *script.rugged*. The verification times using SIS and AQUILA are shown in the last two columns. Because no external don't care is used for optimization, SIS can verify the example with 8-bit data width. However, it fails on larger ones.

7.6 Summary

In this chapter we have introduced two new techniques to extend a gate-to-gate equivalence checker for RTL-to-gate level verification. First, we utilize the external don't care information during the verification process to reduce the possibility of a false negative result. Secondly, we integrate the BDD-based symbolic FSM traversal into the incremental verification framework. This hybrid approach can efficiently extract a subset of unreachable states to speed up the sequential backward justification process in the subsequent incremental verification. This technique is particularly useful for a large design with re-encoded controllers. A reader may also examine Roth's work on verification [113]. His program, VERIFY, preforms logic equivalence checking for two versions of a design. D-algorithm is applied for justification in a miter-like situation. A modification of the D-algorithm called R-algorithm, is used to verify the implementation of an incompletely specified design.

PART II

LOGIC DEBUGGING

Chapter 8

Introduction to Logic Debugging

In this chapter we review a number of representative algorithms for error diagnosis and correction. For error diagnosis, we discuss methods of locating the error sources in an incorrect combinational implementation using BDD techniques or logic simulation. For error correction, we illustrate the techniques of rectifying the circuit by re-synthesizing a number of signals, or by matching the erroneous behavior with a pre-defined error type, (e.g., a missing inverter). Finally, we describe a logic rectification approach using equivalence checking techniques and a heuristic called back-substitution.

8.1 Introduction

Given a gate-level implementation that does not conform to the behavior defined in a specification, error diagnosis is performed to locate the potential error sources. Based on the diagnosis results, error correction can then be performed to fix errors. Here, the specification that serves as a golden reference model could be an RTL description, an arithmetic expression, or a netlist. Because sequential error diagnosis and correction (EDAC) remains a very difficult problem, most existing algorithms target combinational circuits only. EDAC problem often arises in a highly customized design flow where significant amount of human effort is involved. The importance of EDAC tools cannot be over-emphasized because the logic debugging process would be unacceptably time-consuming and frustrating without them.

In addition to fixing implementation errors, the EDAC techniques are also useful in the process of design revision due to specification changes. A specification change is performed when an error in the high-level specification is found during the design validation[1] process. In practice, a specification error could be found in the late design cycle when the logic synthesis or even placement-and-route has been completed. In order to re-use the engineering effort, a designer tends to patch the old implementation and make it conform to the new specification, instead of re-synthesizing from scratch. This process is also known as *engineering change* [130]. Engineering changes allows efficient design revisions.

Both engineering changes and error corrections can be formulated as *rectification problems* in the logic domain. Suppose we have two gate-level netlists, one of which is considered as the specification and the other as the implementation. The objective is to find a transformation for the implementation so that the resulting new implementation is functionally equivalent to the specification. The quality of the rectification can be measured by the amount of logic in the old implementation re-used in the new implementation. Higher *re-use rate* (also called *recycling rate*) indicates a higher quality of rectification.

In this chapter, we discuss representative algorithms for error diagnosis and correction. First, we introduce the notations that will be used throughout this chapter:

- The specification and the implementation are denoted as C_1 and C_2, respectively.
- Both C_1 and C_2 have m primary inputs denoted as $X = \{x_1, x_2, ..., x_m\}$.

1. Design validation determines if the design is *specified correctly*, as opposed to logic verification, which determines if the design is *implemented correctly*.

- The specification C_1 has n primary outputs denoted as $S = \{S_1, S_2, ..., S_n\}$.
- The implementation C_2 also has n primary outputs denoted as $I = \{I_1, I_2, ..., I_n\}$.

Definition 8.1 (S_i, I_i) is called the i-th *primary output pair.*

Definition 8.2 An *erroneous (input) vector* is a binary input vector that can differentiate at least one primary output pair, i.e., v is an erroneous vector if $\exists i \ni (S_i(v) \neq I_i(v))$.

Definition 8.3 $I_i(X, f)$ represents the function of I_i in the network that treats signal f as a pseudo primary input as shown in Fig. 8.1.

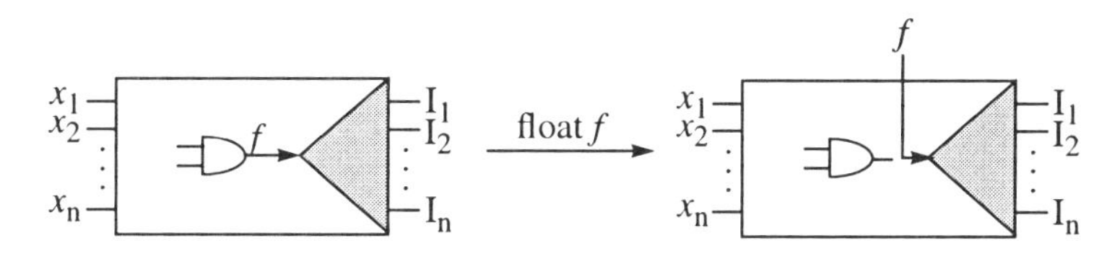

Fig. 8.1 Treating an internal signal as a pseudo primary input.

Definition 8.4 (*Boolean difference*) The Boolean difference of a primary output I_i with respect to a signal f in C_2, denoted as dI_i/df, is defined as:

$$\frac{dI_i}{df} = I_i(X, f = 0) \oplus I_i(X, f = 1)$$

Boolean difference dI_i/df characterizes the set of input vectors that can sensitize a difference at f to the primary output I_i.

8.2 Symbolic Approach

In this section, we discuss the basic idea of using BDDs to automate the EDAC process. More details of this approach can be found in [40,86,87,88,92].

8.2.1 Search for Error Locations

The first step of most EDAC algorithms is to search for the error locations. For simplicity, we first assume that the erroneous implementation can be completely rectified by re-synthesizing an internal signal f as shown in Fig. 8.2. If such a signal, called a *single-fix signal*, exists then the implementation is called *single-signal correctable.* Searching for a single-fix signal can be done precisely using the BDD techniques.

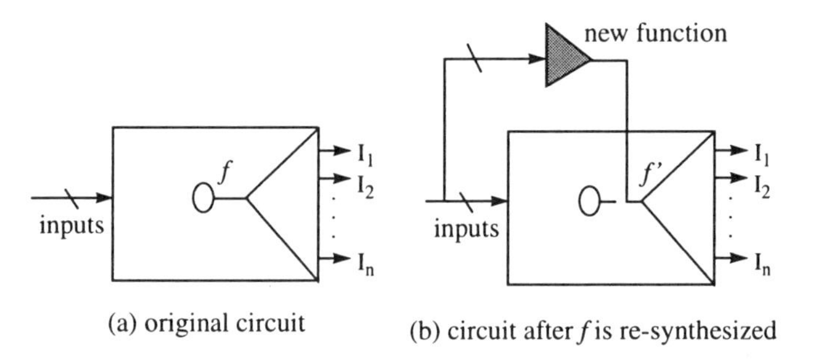

Fig. 8.2 Re-synthesis of an internal signal.

Definition 8.5 (*unconstrained erroneous vector set*) Suppose f is a signal in the implementation C_2, and $I_i(X, f)$ represents the function of output I_i in the network treating f as a pseudo primary input. The *f-unconstrained erroneous vector set* is defined as follows:

$$E(X, f) = \sum_{i=1}^{n} (S_i(X) \oplus I_i(X, f))$$

Function $E(X, f)$ represents the set of vectors defined over $\{x_1, x_2, ..., x_m, f\}$ that causes at least one primary output in the implementation inconsistent with the specification. If f is a single-fix signal, then there should exist a new function at f, denoted as $f^{new}(X)$, so that $E(X, f)$ becomes zero when the variable f is substituted by the new function $f^{new}(X)$. Such a fix function at f does not exist if f is not a single-fix signal.

In the following, we derive the single-fix signal criterion and give a constructive procedure for finding a fix function if one exists.

Single-fix function computation

The Shannon expansion of $E(X, f)$ with respect to f is given by:

$$E(X, f) = \overline{(f)} \cdot E(X, f = 0) + f \cdot E(X, f = 1)$$

We replace the variable f with a new function $f^{\text{new}}(X)$, and denote the result as $E_{\text{constrained}}$:

$$E_{constrained}(X) = \overline{f^{new}(X)} \bullet \underbrace{E(X, 0)}_{\text{on-set}} + f^{new}(X) \bullet \underbrace{E(X, 1)}_{\text{off-set}}$$

The goal is to find a $f^{\text{new}}(X)$ such that $E_{\text{constrained}}(X)$ is a zero function. This $f^{\text{new}}(X)$ should satisfy two conditions so that both product terms of $E_{\text{constrained}}(X)$ are zero functions:

(1) For an input vector $x \in E(X, 0)$, $\overline{f^{new}(x)} = 0$ such that $\overline{f^{new}(x)} \cdot E(x, 0) = 0$.

(2) For an input vector $x \in E(X, 1)$, $f^{new}(x) = 0$ such that $f^{new}(x) \cdot E(x, 1) = 0$.

Therefore, we can conclude that $f^{\text{new}}(X)$ should satisfy the following on-set and off-set constraints to be a single-fix function:

- on-set constraint: $f^{\text{new-on}}(X) = E(X, 0)$ according to condition (1) above.
- off-set constraint $f^{\text{new-off}}(X) = E(X, 1)$ according to condition (2) above.

Any Boolean function that satisfies these two constraints at the same time is a valid fix-function at f. However, if the required on-set and off-set are not disjoint and there exists no such function, then signal f is not a single-fix signal. The following lemma summarizes the above discussion.

Lemma 8.1 [40,114] A signal f in the implementation is a single-fix signal if and only if $E(X, f = 0) \cdot E(X, f = 1) \equiv 0$. Let $f^{new}(X)$ be a Boolean function, then $f^{new}(X)$ is a single-fix function at f if and only if

$$E(X, f = 0) \le f^{new}(X) \le \overline{E(X, f = 1)}$$

8.2.2 Correction by Re-synthesis

Lemma 8.1 points out that a fix function at f, if one exists, is an incompletely specified function with on-set $E(X, f = 0)$, off-set $E(X, f = 1)$, and don't care set $\overline{E(X, f = 0)} \cdot \overline{E(X, f = 1)}$. Assume that $E(X, f = 0)$ and $E(X, f = 1)$ are represented as two BDDs[2]. The procedure of synthesizing such a function consists of three main steps [87]:

Step 1: Minimize the on-set function $E(X, f = 0)$ under the don't care condition of $\overline{E(X, f = 0)} \cdot \overline{E(X, f = 1)}$ using BDD don't-care minimization techniques [58,120]. The resulting BDD represents a fix function $f^{new}(X)$.

Step 2: Convert the BDD of $f^{new}(X)$ to a Boolean network and minimize the network.

Step 3: Replace the signal f with the new minimized network.

8.2.3 Generalization

In many cases, there exists no single-fix signal, especially for the engineering change problem where the implementation usually cannot be fixed by just changing a gate or a wire. The above algorithm for single signal correctable implementation can be further generalized for those situations in at least two different ways: (1) output partitioning and (2) multiple signal re-synthesis.

---

2. Computing the BDD representation of $E(X, f)$ is similar to the procedure of building BDDs for a network except treating $\{x_1, x_2, ..., x_m, f\}$, instead of just $\{x_1, x_2, ..., x_m\}$ as supporting variables.

Output partitioning

An approach based on the single signal correctable condition was proposed for engineering change [87]. In addition to using the above symbolic techniques to find a single-fix signal and its new function, the redundancy-addition-and-removal techniques [27,33] are also known to be effective for increasing the re-cycling rate (i.e., the amount of logic re-used). For a non-single-signal correctable circuit, a heuristic partitions primary outputs into groups such that there exists one single-fix signal for each primary output group. This approach guarantees a solution for any given specification and implementation. However, in the worst case, it may re-synthesize all primary outputs when every primary output forms a distinct group.

Multiple signal re-synthesis

Instead of partitioning the primary outputs into groups, another solution is to fix the entire circuit by re-synthesizing multiple signals simultaneously. For example, if a circuit fails to be single-signal correctable, there may exist two signals (f_1, f_2) that can jointly fix the circuit as in Fig. 8.3.

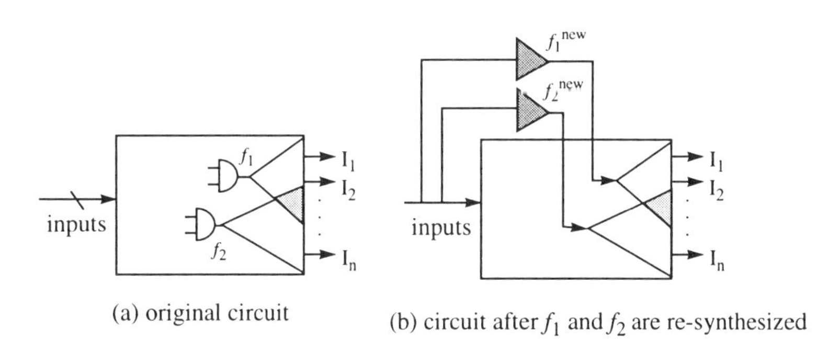

Fig. 8.3 Re-synthesizing two internal signals simultaneously.

This alternative has a much higher computational complexity due to the larger number of candidates that needs to be checked. If the implementation has k signals, then the number of candidate signal pairs is $k(k-1)/2$. For each signal pair (f_1, f_2), the unconstrained erroneous vector set, denoted as $E(X, f_1, f_2)$, is computed by:

$$E(X, f_1, f_2) = \sum_{i=1}^{n} (S_i(X) \oplus I_i(X, f_1, f_2))$$

The new functions f_1^{new} and f_2^{new} that fix the circuit should reduce the E set to zero when substituted into the above formula:

$$E_{constrained}(X) = E(X, f_1^{new}(X), f_2^{new}(X)) = 0$$

The details of solving this boolean equation with two variable functions, f_1^{new} and f_2^{new}, can be found in the literature [48,88,96].

8.3 Simulation-Based Approach

Simulation-based approaches are not as accurate as the symbolic approaches. They try to guess the error locations by simulating a large number of input vectors. The inputs for simulation could be random patterns, test patterns, or distinguishing vectors produced by equivalence checkers. Compared with symbolic approaches, most simulation-based methods provide much less information about how to fix the circuit. Thus an *error model* [1] is usually employed to guide the correction process. An error model proposed by Abadir et al. consists of 10 commonly encountered error types as shown in Fig. 8.4. For a suspected error location, the 10 error types are tried out to see if one of them can match the erroneous behavior of the implementation. This diagnose-and-correct process is repeated until the implementation becomes correct with respect to the entire set of given vectors. Although the correction process is relatively ad-hoc, the diagnosis algorithms using simulation are highly scalable and suitable for real-life designs.

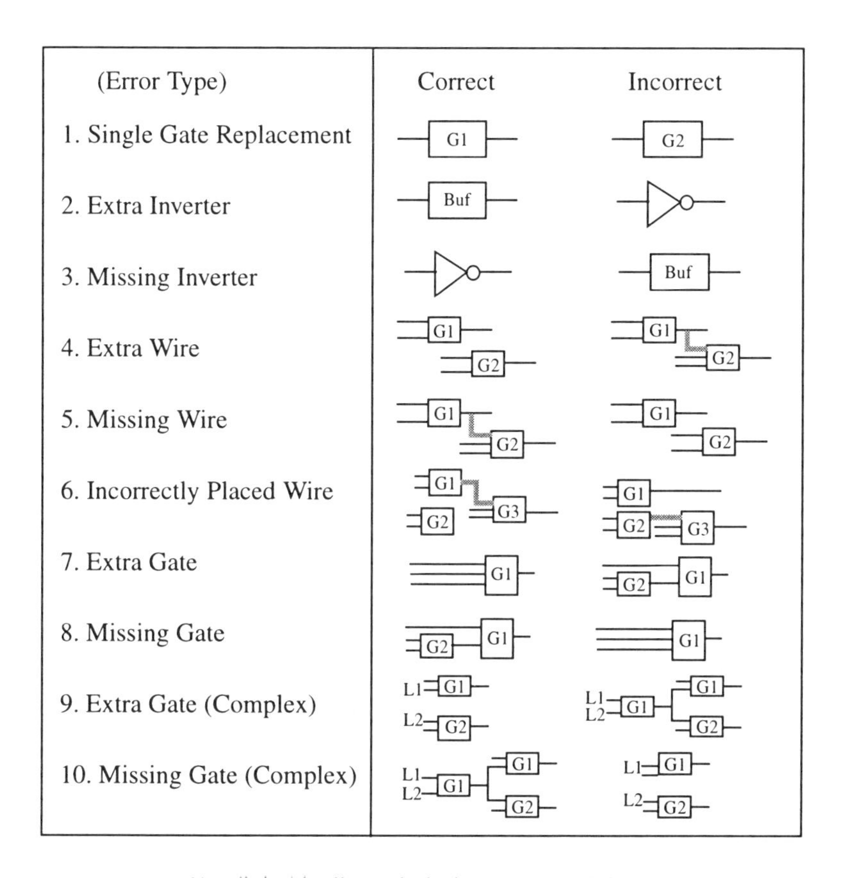

Fig. 8.4 Abadir et al. design error model.

8.3.1 Cone Intersection

The simulation-based approaches can be viewed as filters that eliminate impossible error locations through heuristics. A simple criterion that can be used for filtering is called *cone intersection*. For a given set of input vectors, the primary outputs of the implementation can be divided into two groups, *correct outputs*, and *incorrect outputs* based on simulation results. A signal *f* that is not in the intersection of the incorrect outputs' fanin cones cannot be solely responsible for the error and, thus, cannot be the error site.

Example 8.2 Fig. 8.5 shows the ISCAS-85 benchmark circuit *c*17. The NAND gate at signal *b* is mistakenly implemented as an AND gate, or equivalently, an inverter is missing at signal *b*. Suppose both primary outputs are found erroneous for some input vector. The intersection of the input cones of the incorrect primary outputs are marked in bold lines. Using cone intersection, we conclude that if there is only one error, then signal *b* and *c* are the only possible error sites.

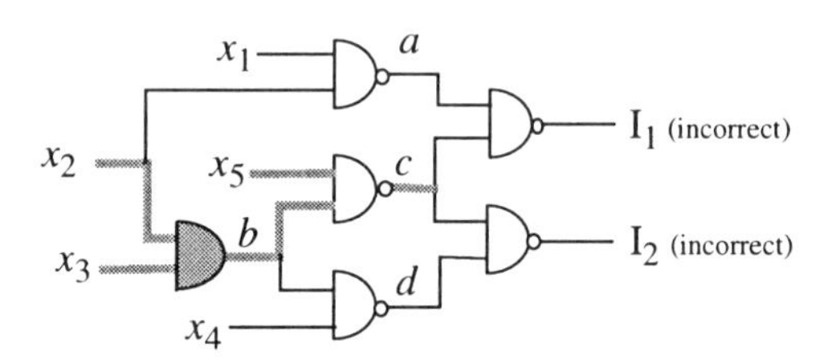

Fig. 8.5 Diagnosis by cone intersection.

Cone intersection is valid under the assumption that there is only one error source in the circuit. The criterion needs to be adapted for circuits with multiple errors. Here, we consider a primary output I_i as reachable from a signal *f* if there exists a topological path from signal *f* to the primary output I_i.

Implication 8.1 Let $\sigma = \{s_1, s_2, ..., s_k\}$ be a set of *k* signals. If the *union* of these *k* signal's reachable primary outputs does not cover *all erroneous primary outputs*, then σ will not be a *k*-error source.

Usually this criterion alone does not yield good diagnosis results. In the following, we describe several improvements that produce more accurate results.

8.3.2 Filter Based on Sensitization

Pomeranz and Reddy proposed a filter for locating the error sites of an erroneous combinational circuit [109]. The idea is based on the observation that if *v* cannot sensitize a discrepancy from a signal *f* to a primary output

I_k, then the erroneous output response of I_k with respect to v cannot be corrected by changing the function of f. In other words, f is *not* responsible for the erroneous I_k with respect to vector v if $v \notin (dI_k/df)$.

```
-by-sensitization(V, C1, C2)
V: a set of erroneous vectors (that can differentiate at least one PO-pair).
C1: specification.
C2: erroneous implementation.
{
    /*--- enumerate every erroneous vector ---*/
    foreach-error-vector v in V {
        perform logic simulation for C1 and C2.

        /*--- enumerate every erroneous PO for the target vector ---*/
        foreach erroneous PO, Ik, in C2 with respect to v {
            S = {every signal in the fanin cone of Ik};
            foreach signal f in S {
                complement the value at f;
                imply the effect-of-change;
                if (the value of Ik is changed){
                    mark f as a candidate error site;
                }
            }
        }
    }
}
```

Fig. 8.6 Finding candidate error sites based on sensitization.

The procedure is shown in Fig. 8.6. It takes a set of erroneous vectors as inputs in addition to the specification and implementation. Note that in this procedure, a signal f could have multiple chances of being marked as a candidate error site because it may have more than one reachable erroneous primary outputs. Checking if a vector can sensitize a discrepancy from f to I_k can be done in three steps:

- Step 1: *Complement* the logic value at f.
- Step 2: Evaluate the *effect-of-change* at f by simulating the fanout cone of f.

- Step 3: If the logic value at l_k is changed, then v can *sensitizes* a discrepancy from f to l_k, Otherwise, it cannot.

Example 8.3 Fig. 8.7 shows the value at each signal of $c17$ in response to the input vector $v = \{(x_1, x_2, x_3, x_4, x_5) \mid (1,1,1,1,1)\}$. Suppose, after comparing with the correct output responses of the specification with respect to this vector, l_2 is found to be erroneous. According to the above algorithm, in addition to the erroneous primary output l_2, only signal b can be regarded as a candidate error site because injecting a '0' at b will create a change at l_2 through the two co-sensitized paths as shown in the figure.

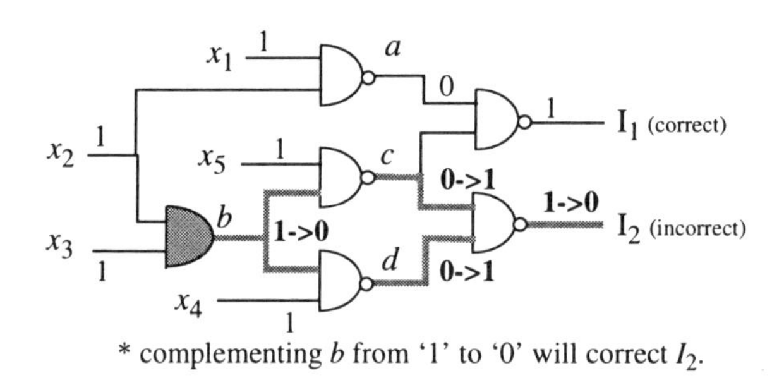

Fig. 8.7 Example of diagnosis by sensitization.

8.3.3 Back Propagation

Kuehlmann et al. proposed another diagnosis algorithm [73]. For a specific erroneous vector, the idea is to determine a potential suspect region by an error tracing process starting from an erroneous output and tracing toward primary inputs. This procedure is a modification of the *critical path tracing* techniques [3]. Fig. 8.8 gives an outline. It consists of two-level loops. The first loop iterates through every erroneous input vector v. The second loop considers one erroneous primary output, l_k, at a time. Each iteration generates a set of suspected signals by invoking a backward propagation routine called *push_back*(l_k). Push_back(), the most important part of this method, is a recursive routine taking a signal as the input. It decides

```
Diagnosis-by-back-propagation (V, C1, C2)
V: a set of erroneous vectors (that can differentiate at least one PO-pair).
C1: specification.
C2: erroneous implementation.
{
    /*--- enumerate every erroneous vector ---*/
    foreach-error-vector v in V {
        perform logic simulation for C1 and C2.

        /*--- push back error effect from each erroneous PO ---*/
        foreach erroneous primary output, Ik, w.r.t. v {
            Push_back(Ik);
            report suspect region for Ik under input vector v;
        }
    }
    report sorted suspect signals according their counter values;
}

Push_back(target signal f)
{
    if (f has been marked suspect) return;
    increment the counter of f by 1; // for ranking suspect signals;
    Identify fanins of f that are regarded as suspects by propagation rules;
    foreach suspect fanin w {
        Push_back(w); // only further propagate towards suspect fanin
    }
}
```

Fig. 8.8 Diagnosis by back propagation.

which fanin(s) of the target gate at f should be held suspect in addition to the target signal f. This is done by examining the logic values at the target signal f and each of its fanins with respect to the erroneous vector under consideration using a set of rules.

The propagation rules [73] are defined for some primitive gate such as AND, OR, and INVERTER. A complex gate needs to be decomposed into primitive gates first before applying these rules. Consider an AND-gate f. Depending on the logic value at f, there can be two cases:

(1) The controlling case ('0' for AND-gate): only fanins of f with controlling values ('0') are considered as suspects. The idea is that if

f is not correct, inverting a fanin from non-controlling to controlling value does not cause a change at f (i.e., f stays controlling). Because of this, non-controlling fanins should not be held responsible for the error.

(2) The non-controlling case ('1' for AND-gate): every fanin is considered a suspect, because inverting any one of them from the non-controlling value to the controlling value suffices to change the value of f.

Example 8.4 Fig. 8.9 shows the value at each signal of $c17$ in response to the input vector $v = \{(x_1, x_2, x_3, x_4, x_5) \mid (1,1,1,1,0)\}$. Suppose both primary outputs are found to be erroneous. Applying the back propagation algorithm, the suspect regions for the two POs are $\{c, b\}$ and $\{c, b, d\}$, respectively. The signals traversed from the two POs are marked in bold lines in Fig. 8.9(a) and Fig. 8.9(b), respectively. In comparison, back propagation is not as accurate as a sensitization-based technique, which will produce suspect regions $\{c, b\}$ and $\{b\}$ for the two erroneous POs, respectively. However, back propagation only needs to sweep the implementation from POs to PIs once for each erroneous vector and, thus, is more efficient than the sensitization-based methods. In many cases, it produces quite satisfactory results [73].

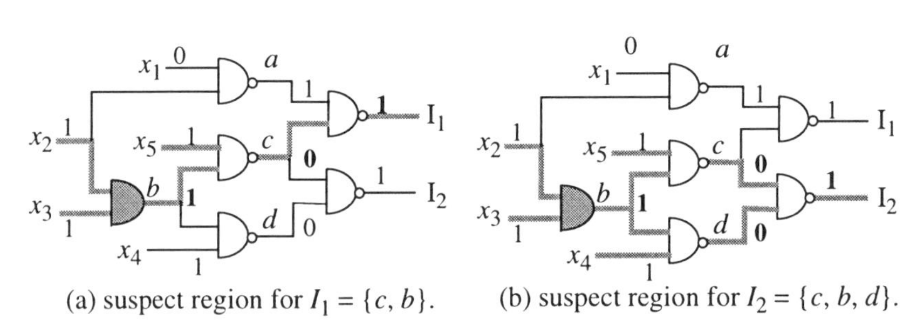

(a) suspect region for $I_1 = \{c, b\}$. (b) suspect region for $I_2 = \{c, b, d\}$.

Fig. 8.9 Example of diagnosis by back propagation.

In a circuits with multiple errors, designers tend to find-and-fix one error at a time. Thus it is often desirable to rank the suspect signals during the diagnosis process so that a highly suspicious signal can be examined first for manual or automatic correction in the iterative process. The above algorithm can be easily modified to estimate the "error probability" of each suspect signal for this purpose. For example, a counter can be used for each signal. Whenever a signal is included in a suspect region during the back propagation process, the counter is incremented by 1. After examining all vectors, the suspect signals are sorted according to their respective counter values. A signal with a higher count is considered as a more suspicious error signal.

8.3.4 Enhancement with Observability Measure

The accuracy of back propagation can be further improved without sacrificing the efficiency [126,127]. One improvement technique assigns a weight to each signal, called *observability measure* (OM) [126]. The observability measure indicates the potential of a signal to be a suspect. Consider an NAND-gate whose output signal f is being processed during the back propagation under an input vector v with respect to erroneous primary output I_k. Fig. 8.10(a) shows the case when the logic value of f is the controlling value. The observability measure of signal f, assumed to be w, is *split evenly among the fanin signals with controlling value*. In this

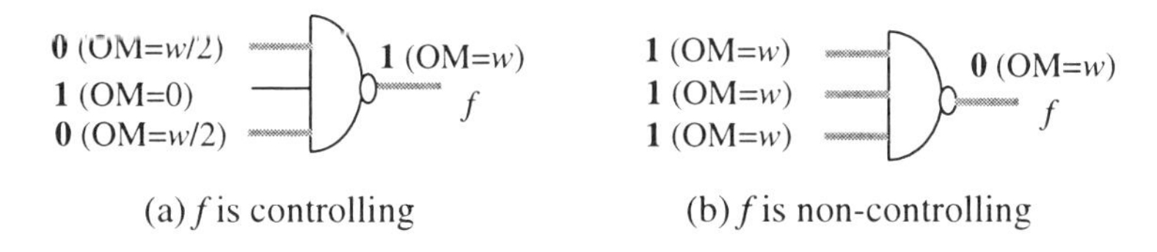

Fig. 8.10 Enhanced back propagation rule with observability measure.

example, there are two such fanins and, hence, each of them is assigned an observability measure of $w/2$. Fig. 8.10(b) shows the case when the logic value of f is the non-controlling value. Then the observability measure at the gate's output is *copied to each fanin signal*. In this example, there are three fanins, each of which is assigned an observability measure of w.

Finally, we consider a stem signal with multiple fanout branches. As a heuristic, the observability measure of such a signal is defined as the smaller of 1 and the summation of its branch observability measures as shown in Fig. 8.11.

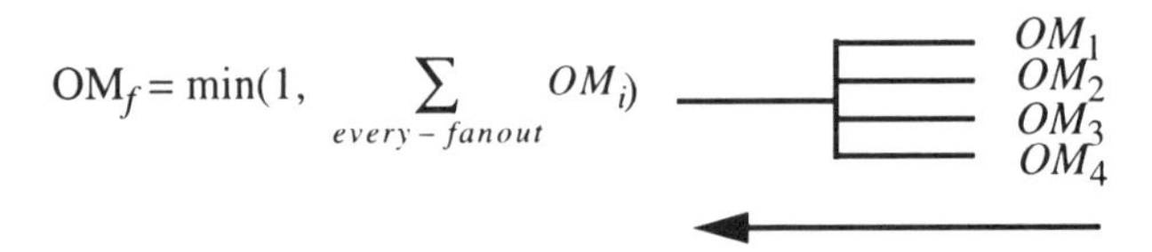

Fig. 8.11 Backward propagation of observability measure at a stem signal.

It can be shown [126] that if a signal f has an observability measure less than 1, then the input vector v under consideration is not in the Boolean difference of the output I_k under consideration with respect to f, i.e., $v \notin (dI_k/df)$. Hence, the incorrect response of I_k with respect to v cannot be fixed by re-synthesizing this signal alone, and thus, it is safe to exclude f from the suspect signal list. In other words, after the back propagation process, only signals with an observability measure 1 are in the suspect region. For Example 8.4, this technique can further reduce the suspect region of the second output with respect to input vector $v = \{(x_1, x_2, x_3, x_4, x_5) \mid (1,1,1,1,0)\}$ from $\{b, c, d\}$ to $\{b\}$.

8.4 Structural Approach

An algorithm based on structural information and equivalence checking techniques was proposed for the purpose of engineering change [14]. The approach relies on a sequence of *normal substitutions* and *back-substitutions* to incrementally rectify the incorrect implementation. The normal substitution is an operation that replaces a signal in the specification by a *permissible* signal in the implementation. Fig. 8.12(a) shows an example. The specification and implementation only differ in one gate at signal d_1 of

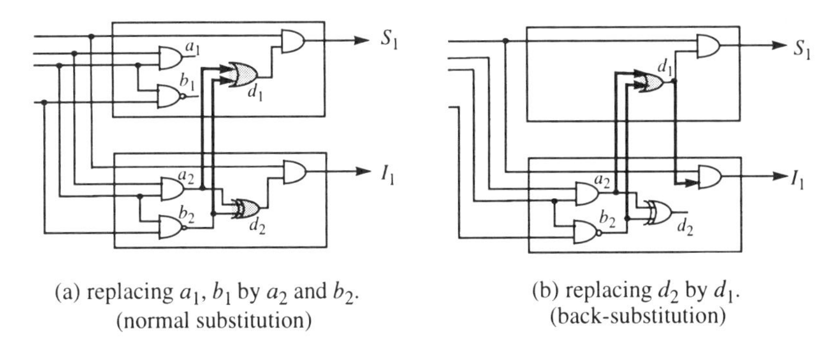

(a) replacing a_1, b_1 by a_2 and b_2. (normal substitution)

(b) replacing d_2 by d_1. (back-substitution)

Fig. 8.12 Normal substitution and back-substitution.

the specification and d_2 of the implementation. By applying the incremental verification techniques described in Chapter 3, signal pairs (a_1, a_2) and (b_1, b_2) can be proven as functionally equivalent. Thus, signals a_1 and a_2 can be replaced by signals b_1 and b_2, respectively, without changing the function of the specification. A normal substitution can be viewed as a *recycling operation* - which utilizes the existing logic in the implementation to realize the internal functions of the specification. As a result, it contributes to better rectification results.

On the other hand, back-substitution is an operation that replaces a signal in the implementation with a signal in the specification as shown in Fig. 8.12(b). Note that (d_1, d_2) are not equivalent in this example and, thus, this back-substitution will change the function of the implementation. Back substitution is a heuristic that tries to reduce the difference between the specification and the implementation. To be effective, the back-substitution transformation should be performed at the suspicious error locations of the implementation, and guided by some structural correspondence between the specification and the implementation. The objective of a back-substitution transformation is to create more equivalent signal pairs in the fanout cones of the location where back-substitution is performed. In the example of Fig. 8.12, signal d_2 is indeed a suspect error location, and d_1 is the structurally corresponding signal of d_2 (this may have to be specified by the user). Replacing d_2 by d_1 successfully corrects the primary output function of C_2. The procedure of using normal substitution and back-substitution for logic rectification is summarized in Fig. 8.13. Once the joined network that connects the primary inputs of the two circuits together is constructed, the process iterates until the function of every primary output pair of the specification and the implementation become the same. Each iteration may consist of a sequence of normal substitutions followed by one back-substitution as shown in the figure.

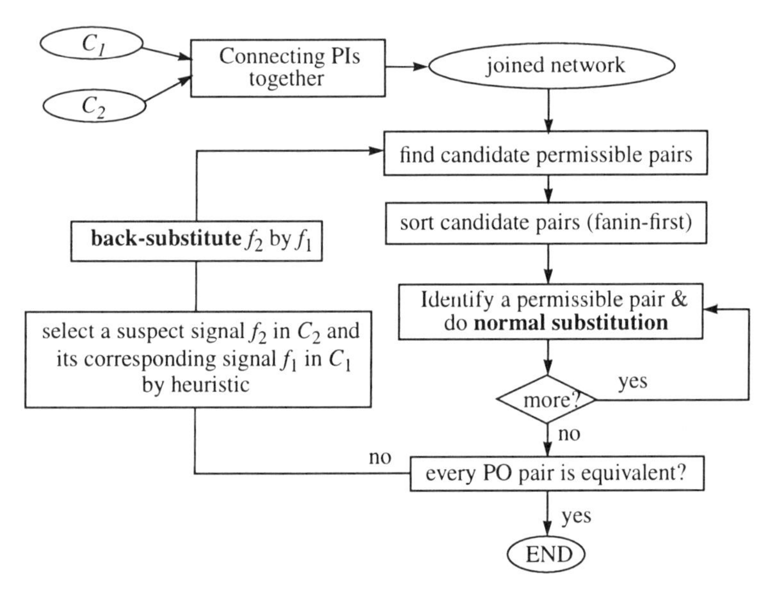

Fig. 8.13 Flowchart of the incremental logic rectification.

8.5 Summary

Symbolic design error diagnosis and correction have been studied for a long time. With the advent of the BDD techniques, these classic ideas have become more attractive and were revisited by several research groups. Not only can symbolic approaches pin-point the exact error locations, they may also suggest a way to fix the design. On the other hand, simulation-based approaches are primarily filters, which gradually exclude signals from the candidate list of error locations. In general, the simulation-based approaches are much more scalable than the symbolic ones. In many cases they also produce satisfactory diagnosis results. However, the simulation-based approaches do not provide enough insight about how to fix the error(s). Thus, they are less suitable for applications that require automatic correction (e.g., the engineering change problem). In addition to the symbolic approaches and simulation-based approaches, several other techniques such as redundancy-addition-and-removal and logic equivalence checking have also been shown to be useful in further improving the quality of error diagnosis and/or correction.

Chapter 9

ErrorTracer: Error Diagnosis by Fault Simulation

In this chapter we discuss a method for combinational design error diagnosis. We introduce a fault-simulation-based technique to approximate each signal's correcting power. The correcting power of a signal is measured in terms of the size of the signal's *correctable set*, namely, the maximum set of erroneous input vectors that can be corrected by re-synthesizing the signal. Only the signals that can correct every erroneous input vector are considered as potential error sources.

9.1 Introduction

In this chapter, error diagnosis is formulated as a fault simulation process using a number of erroneous vectors as simulation vectors, where an erroneous vector is an input vector that can differentiate the specification and the implementation. The erroneous vectors are derived in advance by simulating random vectors or by using an equivalence checker. This approach is based on a notion called *correctable set*. A signal's correctable set is defined as the set of erroneous vectors that can be corrected by re-synthesizing the signal. Let v be an erroneous vector and f be a signal in the erroneous circuit. It can be shown that whether v is correctable by f can be *precisely* determined by simulating input vector v for the stuck-at faults at f. Similar to most simulation-based approaches, this algorithm is a monotone filtering process. Initially every signal is considered as a candidate. Every erroneous vector is simulated for the stuck-at faults at each candidate signal. According to the fault simulation results, if a signal is unable to correct any erroneous vector, then it is a false single-fix candidate. False candidates are removed immediately from the candidate list before simulating the next erroneous vector.

Theoretically speaking, this approach could be as accurate as the symbolic approaches if the entire erroneous vector set is simulated. The experimental results show that this method is also very efficient for single-signal correctable circuits. This approach can be further generalized for circuits with multiple errors. For diagnosing multiple errors, this algorithm searches for a set of signals that can jointly fix the erroneous circuit. Because the number of candidate sets of signals grows rapidly with the cardinality of the multiple errors, a two-stage fault simulation procedure is used to speed up the process. This two-stage procedure, utilizing the topological dominance relation between signals, does not cause any loss of accuracy. The results of diagnosing double errors using this approach for all ISCAS-85 benchmark circuits are reported in Section 9.5. For the larger ones in this benchmark set, the BDD-based symbolic approaches fail to complete.

The rest of this chapter is organized as follows. Section 9.2 gives the basic assumptions and definitions. Section 9.3 describes the algorithm for single-signal correctable circuits. In Section 9.4, it is generalized for diagnosing multiple errors. Section 9.5 presents some experimental results.

9.2 Basic Terminology

We assume that the specification and the erroneous implementation are given in the gate-level. Both share the same set of primary inputs, denoted as $\{x_1, x_2, ..., x_n\}$. The primary outputs of the specification and the implementation are denoted as $\{S_1, S_2, ..., S_m\}$ and $\{I_1, I_2, ..., I_m\}$, respectively.

Definition 9.1 (*erroneous output*) If an erroneous vector v can differentiate the i-th primary output pair, then I_i is an erroneous output; otherwise, I_i is a correct output with respect to v. Given an erroneous vector v, we can partition the primary outputs of the implementation, C_2, into two groups with respect to that vector: (1) erroneous output group, *Error_PO*(v), and (2) correct output group, *Correct_PO*(v).

Definition 9.2 (*sensitization set*) For a signal f and a primary output I_i in C_2, the sensitization set, denoted as $SEN_i(f)$, is the set of input vectors that can sensitize a discrepancy from f to I_i. Boolean difference dI_i/df is the characteristic function of the sensitization set $SEN_i(f)$. $SEN_i(f)$ represents those input vectors for which signal f determines the value at I_i.

9.3 Single Error Diagnosis

We first introduce the notion of the correctable vector. The necessary and sufficient condition for single-fix signal is then presented based on a slightly different point of view than that used before [40,92]. Finally, the overall algorithm is described.

9.3.1 Correctability

Definition 9.3 (*correctable vector*) An erroneous vector v is *correctable* by a signal f in C_2 if there exists a new function for signal f such that v is *not* an erroneous vector for the resulting new circuit.

Proposition 9.1 Let v be an erroneous vector and f be a signal in C_2. Then v is correctable by f if and only if two conditions are satisfied:

(1) v *can* sensitize a discrepancy from f to every erroneous primary output of C_2, i.e., for every primary output I_i in Error_PO(v), $v \in SEN_i(f)$.

(2) v *cannot* sensitize a discrepancy from f to any correct primary output of C_2, i.e., for every primary output I_i in Correct_PO(v), $v \notin SEN_i(f)$.

Proof: Let F be a Boolean function that disagrees with the original function *f* only for the input vector *v*. Then after replacing signal *f* with the new function F, a discrepancy is injected at *f*. If the two above conditions are satisfied, then the response of every erroneous output toggles and thus becomes correct (because *v can* sensitize a discrepancy from *f* to every one of them). At the same time, every originally correct output remains correct (because vector *v cannot* sensitize a discrepancy from *f* to any one of them). On the other hand, it can be shown that if the two conditions are not satisfied, then at least one erroneous primary output will remain erroneous, or at least one originally correct output will become erroneous, regardless of what the new function is. (Q.E.D.)

Proposition 9.2 Signal *f* is a single-fix for C_2 if and only if every erroneous vector is correctable by *f*.

Proof: Here, we make no distinction between a signal and its function. Let $F = f \cdot E' + f' \cdot E$, where E is the characteristic function of the entire erroneous vector set. Intuitively, F can be interpreted as a function that *agrees* with *f* on all non-erroneous vectors, while it *disagrees* on all erroneous vectors. It can be shown that F is a fix function because it satisfies the construction rule of the fix function for each erroneous vector described in the proof of Proposition 9.1. On the other hand, if any erroneous vector cannot be corrected by *f*, then it follows from Definition 9.3 that there exists no a fix function, and *f* is not a single-fix for C_2. (Q.E.D.)

9.3.2 The Algorithm for Single Error Diagnosis

In order to handle large circuits, the identification of the single-fix signals does not rely on the construction of BDDs. Instead, an iterative filtering process is employed to gradually reduce the number of single-fix candidate signals. The overall flow is illustrated in Fig. 9.1.

Simulation using randomly generated input vectors is first preformed to collect a number of erroneous vectors. At the beginning, every signal in C_2 is considered as a single-fix candidate. Then the algorithm starts a two-level loop. The outer loop enumerates every erroneous vector. The inner loop iterates through every single-fix candidate signal. For each erroneous

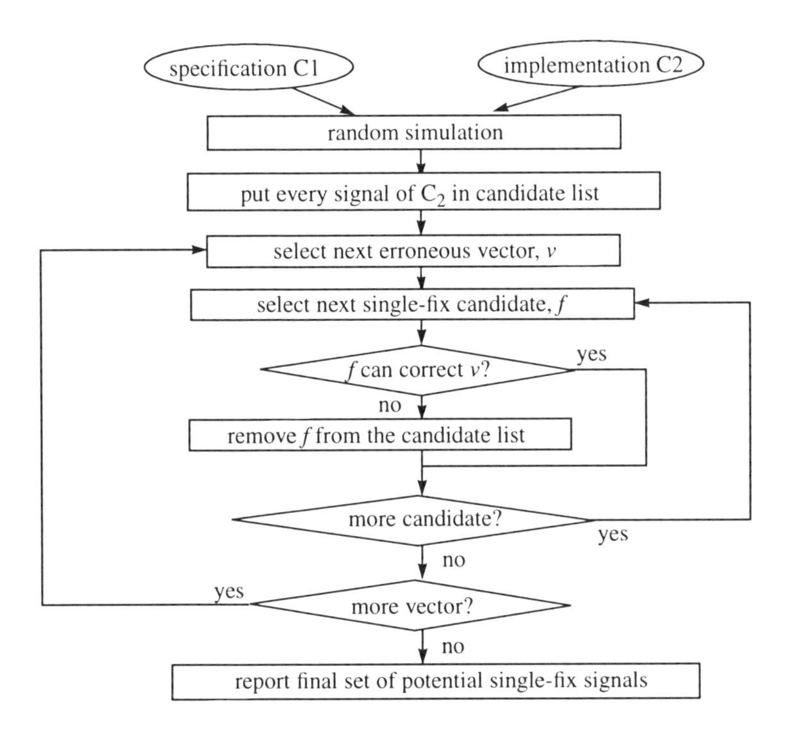

Fig. 9.1 Algorithm for single-signal correctable circuits.

vector v and a target candidate f, it is examined if v is correctable by f via fault simulation (explained later). If signal f cannot correct vector v, then f is not a single-fix signal and is removed from the candidate list. After we have examined every erroneous vector and eliminated false single-fix candidates, a set of potential single-fix signals is derived.

9.3.3 Correctability Check via Fault Simulation

Given an erroneous vector v and a signal f, *correctability check* decides whether v is correctable by f. It is an operation to verify two conditions: (1) if v *can* sensitize a discrepancy from f to *every* erroneous output, and (2) if v *cannot* sensitize a discrepancy from f to any correct output. Proposition 9.3

shows that this can be checked by simulating vector *v* for the stuck-at-0 and stuck-at-1 faults at *f*. In the following discussion, we use $C_{2\,|\,f=0}$ to represent the faulty implementation of C_2 with signal *f* stuck-at-0 fault. Similarly, $C_{2\,|\,f=1}$ represents the faulty implementation of C_2 with signal *f* stuck-at-1 fault.

Proposition 9.3 Signal *f* can correct an erroneous vector *v* if and only if the faulty circuit $C_{2\,|\,f=0}$ or $C_{2\,|\,f=1}$ has the same output response as the specification C_1 with respect to input vector *v*.

Proof: It is obvious that if $C_{2\,|\,f=0}$ or $C_{2\,|\,f=1}$ is equivalent to C_1 with respect to input vector *v*, then *v* is correctable. On the other hand, if neither $C_{2\,|\,f=0}$ nor $C_{2\,|\,f=1}$ is equivalent to C_1 with respect to input vector *v*, then it can be shown that at least one of the two criteria of Proposition 9.1 cannot be satisfied, and thus, there exists no new function for signal *f* to correct *v*. (Q.E.D.)

Based on Proposition 9.3, we can simulate the pre-generated erroneous input vector set for stuck-at faults at each candidate signal to prune out the false single-fix candidates. This process can be further sped up by exploring the topological dominance relation between signals. Let *f* and *dom* be two signals, and signal *dom* be a topological dominator of signal *f*. In other words, every path originated from *f* to any primary output passes through signal *dom*. It has been proven [86] that if *dom* cannot correct an erroneous vector *v*, then *f* cannot correct it, either. Therefore, once a false single-fix candidate is found, we can immediately remove its dominated signals from the candidate list. Fig. 9.2 shows the revised routine of simulating one erroneous vector for correctability check.

First, the candidate list is sorted in a fanout-first order (i.e., every signal is placed after its transitive fanout signals). Given an erroneous vector, the signals are examined sequentially. Then fault simulation is performed for the stuck-at-0 and stuck-at-1 faults for each target signal, *f*. The simulation results are compared with pre-stored output responses of the specification to decide the correctability. If the target signal *f* fails to correct *v*, then the signal *f* along with the signal dominated by *f* are removed from the candidate list. The correctability check iterates until every signal remaining in the candidate list has been checked. In this revised routine, some

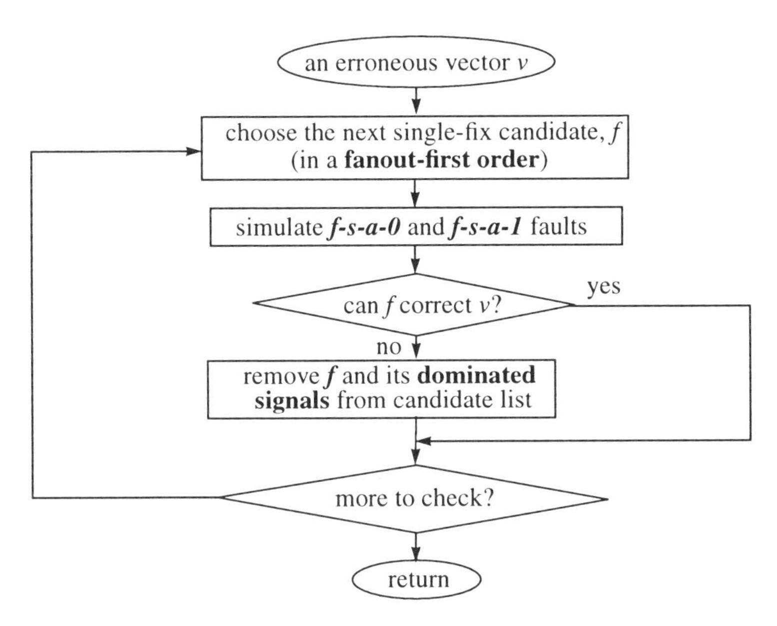

Fig. 9.2 Checking correctability for an erroneous vector.

candidate signals may be dropped without fault simulation because one of their dominators has been proven unable to correct the given erroneous vector.

9.4 Multiple Error Diagnosis

If the algorithm outlined in Fig. 9.1 and Fig. 9.2 proves that there is no single-fix signal, the search can be extended for multiple signals that jointly correct the implementation. In this section the fault simulation-based approach is generalized for circuits with multiple errors. We define the *k*-correctable vector and the *k*-correctable circuit. Then we show how to perform the *k*-correctability check via fault simulation in a two-stage algorithm.

9.4.1 *k*-Correctability

Definition 9.4 (*k-correctable vector*) An erroneous vector v is k-correctable by a set of k signals in C_2, $\sigma = \{f_1, f_2, \ldots, f_k\}$, if there exists a new function for each signal f_i in σ such that v is not an erroneous vector for the resulting new circuit.

Definition 9.5 (*k-correctable circuit*) If the implementation C_2 can be completely corrected by re-synthesizing a set of k signals, $\sigma = \{f_1, f_2, \ldots, f_k\}$, then C_2 is k-correctable and σ is called a *fix set*.

For a given signal set σ, *an enumeration over* σ is an assignment that assigns a binary value to each signal in σ. For a signal set with k signals, there will be 2^k different enumerations, and each enumeration corresponds to a faulty circuit with multiple stuck-at faults. For example, consider a set of two signals $\sigma = \{f_1, f_2\}$. There will be four different enumerations, namely, $\{f_1{=}0, f_2{=}0\}$, $\{f_1{=}0, f_2{=}1\}$, $\{f_1{=}1, f_2{=}0\}$, $\{f_1{=}1, f_2{=}1\}$. Among them, $\pi = \{f_1{-}0, f_2{=}0\}$ defines a faulty circuit with a double-fault (f_1 stuck-at-0 and f_2 stuck-at-0). Proposition 9.4 shows that in order to decide whether an erroneous vector is k-correctable by a set of signals σ, fault simulation needs to be performed on every one of the 2^k faulty circuits defined over σ.

Proposition 9.4 Let σ be a set of k signals and π_i be one of 2^k enumerations defined over σ. Erroneous vector v is k-correctable by σ if and only if at least one of the 2^k faulty circuits $C_{2\,|\,\pi i}$, $i = 1, 2, \ldots, 2^k$, has the same output response as the specification C_1 with respect to the input vector v, where $C_{2\,|\,\pi i}$ denotes the faulty implementation defined by the enumeration π_i.

Proof: Omitted.

The complexity of diagnosing circuits with multiple errors grows polynomially with the number of signals in C_2. For example, to find the signal pair that can jointly fix the circuit may need to examine $p(p\text{-}1)/2$ possible candidate pairs, where p is the number of signals in C_2. Similar to the case of finding the single-fix signals, the topological dominance relation is useful in reducing the complexity. To explain this speedup technique, a dominance relation between two sets of signals is defined.

Definition 9.6 (*set dominance relation*) Let $\sigma_1 = \{f_1, f_2, \ldots, f_k\}$ and $\sigma_2 = \{d_1, d_2, \ldots, d_k\}$ are two sets of signals. If f_i is topologically dominated by d_i for $1 \le i \le k$, then σ_1 *is dominated by* σ_2, denoted as $\sigma_1 < \sigma_2$. We refer to σ_1 as a subordinate set of σ_2.

Proposition 9.5 Let $\sigma_1 = \{f_1, f_2, \ldots, f_k\}$, $\sigma_2 = \{d_1, d_2, \ldots, d_k\}$, and $\sigma_1 < \sigma_2$. If σ_2 cannot correct an erroneous vector v, then σ_1 cannot correct v either.

Proof: Omitted.

9.4.2 A Two-Stage Algorithm for Diagnosing Multiple Errors

Based on the above proposition, we introduce a two-stage, fault simulation-based algorithm for diagnosing multiple-errors. For the sake of simplicity, the discussion is based on the case of $k = 2$. That is, the implementation C_2 is double-signal correctable, but not single-signal correctable. The overall algorithm is illustrated in Fig. 9.3. A signal d in C_2 is referred to as a *key signal* if d is not dominated by any other signal. Signal pair $\{d_1, d_2\}$ is a key signal pair if d_1 and d_2 are both key signals. In the first stage, only key signal pairs are considered as candidates. For an erroneous vector v and a candidate key signal pair $\sigma = \{d_1, d_2\}$, fault simulation for each of the four possible faulty circuits $C_{2 \mid \{d1=0,\ d2=0\}}$, $C_{2 \mid \{d1=0,\ d2=1\}}$, $C_{2 \mid \{d1=1,\ d2=0\}}$, and $C_{2 \mid \{d1=1,\ d2=1\}}$ is performed. If any one of these four faulty circuit has the same output responses as the specification with respect to v, then v is 2-correctable by σ. Otherwise, σ is a false candidate pair and is removed from the candidate list. After the fault simulation process has iterated through every erroneous vector and candidate pair, survivor key signal pairs that have the potential to be the fix pairs are obtained. However, these pairs are only a subset of the potential fix pairs. Every subordinate pair of each of them is also a possible fix pair. For example, suppose $\sigma = \{d_1, d_2\}$ is a potential fix pair after the first stage. Let $\{f\}$ and $\{g_1, g_2\}$ be the sets of signals dominated by d_1 and d_2, respectively. Then $\{(d_1, g_1), (d_1, g_2), (f, g_1), (f, g_2), (f, d_2)\}$ are possible fix pairs too.

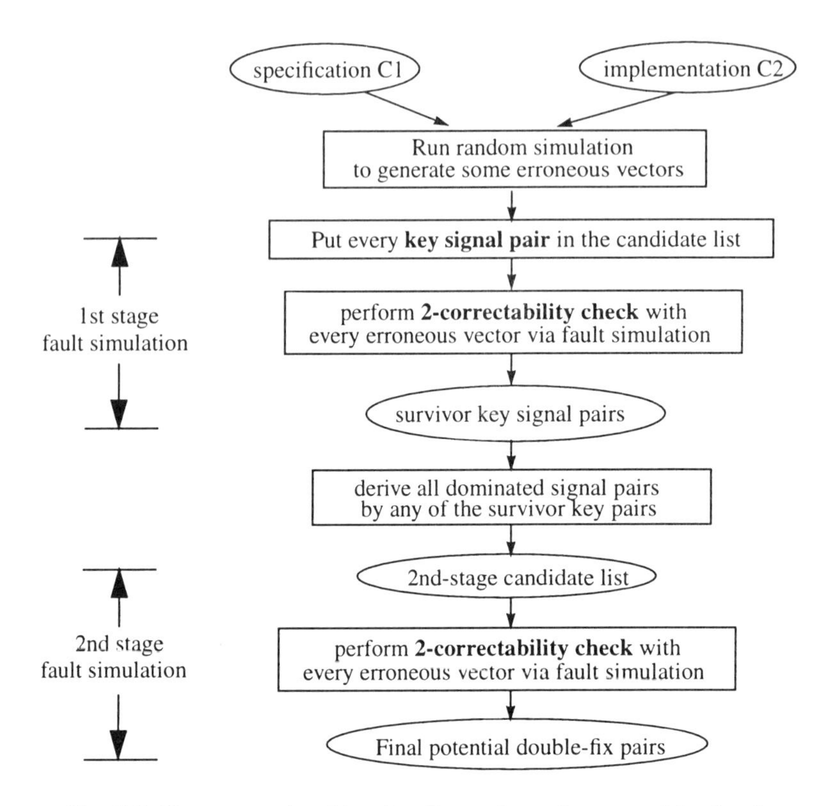

Fig. 9.3 Two-stage algorithm for diagnosing a 2-correctable circuit.

While they are not examined in the first stage. Note that only the subordinate pairs of the survivor key signal pairs need to be further checked in the second stage. The subordinate pairs of those false key signal pairs filtered out in the first stage can never be fix pairs. Usually, the number of subordinate pairs to be checked in the second stage is very small as will be shown in Section 9.5.

9.5 Experimental Results

The algorithm has been implemented in C language in the SIS environment [121]. The program is named *ErrorTracer*. The experiments are performed for the entire set of ISCAS-85 benchmark circuits. Each circuit is first optimized by the optimization script *script.rugged* to obtain an implementation. Then they are decomposed into AND/OR gates using SIS command "*tech_decomp -a 5 -o 5*". To generate erroneous implementations, gate type errors are injected randomly using a program from [73]. This program randomly selects some nodes and then scrambles their functions. Note that this diagnosis approach is based on the notion of re-synthesis and, thus, not restricted to the types of errors introduced in the implementations.

Table 9.1 shows the results of single-error diagnosis. The program was run 20 times for each benchmark circuit: each run diagnosed a different single-error implementation. In the pre-processing stage, random vectors are simulation until 320 erroneous vectors are collected, or 16,000 random patterns have been simulated. For very few cases, no erroneous vector is found after simulating 16,000 random patterns. For these cases, we use the

Table 9.1 Results of diagnosing optimized ISCAS85 benchmark circuits injected with **one** error.

circuit	# signals in C2	**# potential fix signals**	pessimistic lower bound	hit ratio	suspect ratio	CPU-time (seconds)		
						random simulation	fault simulation	total
C432	175	3.4	2.3	100%	1.94%	4	1	5
C499	410	4.7	2.3	100%	1.15%	11	1	12
C880	292	5.2	2.7	100%	1.78%	6	1	7
C1355	410	5.1	2.5	100%	1.32%	15	1	16
C1908	357	4.3	2.4	100%	1.20%	15	2	17
C2670	803	10.6	4.3	100%	1.32%	38	5	43
C3540	1189	7.3	2.6	100%	0.61%	54	6	60
C5315	1248	4.7	2.7	100%	0.38%	76	8	84
C6288	2339	2.9	1.8	100%	0.13%	27	33	60
C7552	2187	18.0	3.4	100%	0.82%	185	20	205
Average	941	6.6	2.7	100%	1.06%	43	8	51

formal equivalence checker, AQUILA, as described in Chapter 5, to determine if the error-injected implementation is functionally equivalent to the specification. For all of these cases, the implementations are equivalent to the specifications. Table 9.1 shows the average results for the real erroneous implementations. Column "*# potential fix signals*" is the number of potential single-fix signals delivered by this program. On an average, the algorithm outputs 6.6 potential single-fix signals for 10 ISCAS-85 benchmark circuits.

Fig. 9.4 shows a graph of the number of potential single-fix signals versus the number of erroneous vectors simulated during the diagnosis process for a single-error implementation of C6288. Initially, every signal

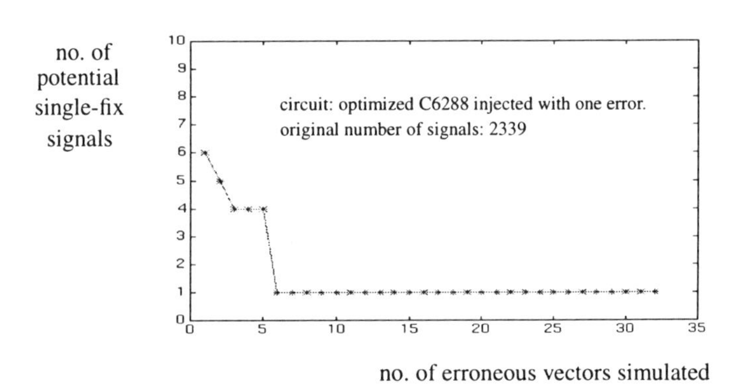

Fig. 9.4 The curve of the accuracy versus the number of erroneous vectors that are fault simulated.

is regarded as a potential single-fix signal, and there are a total of 2339 candidates. This approach narrows down the error region to only 6 signals after simulating only one erroneous vector. After simulating 5 more erroneous vectors, this approach precisely pin-point the location of the injected error. This curve indicates that the conditions for a signal to be a potential single-fix signal used in this approach is very stringent and, thus, the iterative process is able to filter out most false candidates rapidly. In these experiments, the numbers of potential single-fix signals for most cases reduce quickly and stabilize after simulating less than 10 erroneous vectors.

Fig. 9.5 shows the numbers of potential single-fix signals for 20 samples of C6288 and C7552. The numbers of C7552 vary from one erroneous

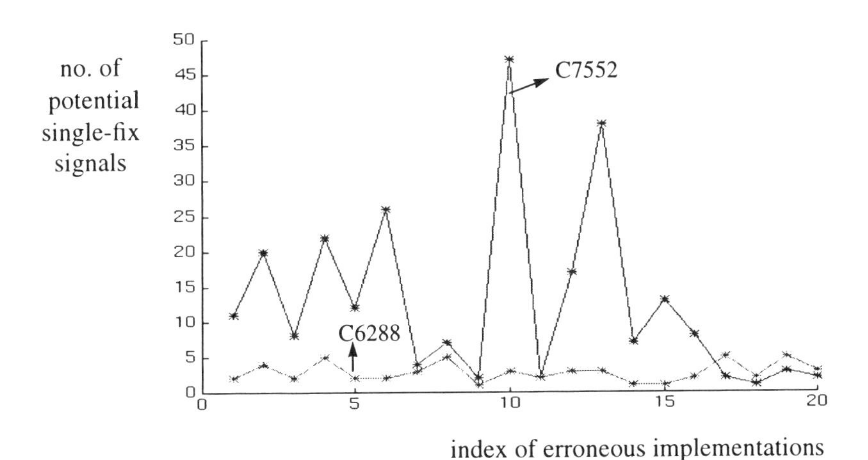

Fig. 9.5 The number of potential fix signals for C6288 and C7552.

implementation to another, while the numbers for C6288 are similar and small. This may indicate that C6288, a 16-bit multiplier, is easier to diagnose because most of its internal signals are multiple fanout stems, and a design error tends to produce many erroneous primary outputs. Column "*Lower bound*" of Table 9.1 is a *pessimistic* lower bound on the total number of single-fix signals. It is obtained by counting the number of dominators of the injected error signal plus one. This is a conservative estimate because some single-fix signals may not dominate the injected error signal and, thus, are not included. Column "*Hit ratio*" indicates the probability that the injected error signal is included in the delivered set of potential fix signals. This approach guarantees a 100% hit ratio. Column "*Suspect ratio*" is the ratio of the number of potential fix singles to the total number of signals in C_2. The average is 1.06%, which means almost 99% of the signals are disqualified as single-fix signals. CPU-time is for 150MHz Sparc20. The total time consists of the random simulation time and the fault simulation time. The random simulation time for generating 320 erroneous vectors (average 40 seconds) dominates the fault simulation time (average 8 seconds). The results also indicate that this program achieves the same level of accuracy by simulating only 32 erroneous patterns for all cases.

Table 9.2 shows the results of diagnosing the implementations injected with two random errors. Because this is a more time-consuming process, only one erroneous implementation is diagnosed for each benchmark circuit. Each erroneous implementation is proven not to be single-signal correctable by the program (i.e., every signal is disqualified as a single-fix). Note that if an implementation is single-signal correctable, then the number of potential double-fix pairs would be huge. For example, if f is a single-fix, then the signal pair (f, x), where x is any signal in C_2, would be a double-fix pair. Due to the large number of the double fix pairs, the diagnosis process could be very time-consuming.

Column "*# candidate pairs checked*" is the number of the candidate pairs checked in the two stages. Usually the number of candidates in the second stage is negligible as compared to the one in the first stage. The ratio of the total number of double-signal pairs to the number of pairs being checked reflects the speedup factor due to the use of the set dominance relation. Consider C432 for example. The number of signals in C_2 is 175, hence the total number of candidate pairs without using any dominance relation will be (175)(175-1)/2 = 15,225. In this approach, the total number of candidate pairs being checked is reduced to only (1035+13) = 1048. Therefore, the speedup is 15225 / 1048 = 14.5. The *hit ratio* is also guaranteed to be 100% for double-error diagnosis. The CPU-time is mostly dominated by the fault simulation time. It is proportional to the number of candidate pairs. C6288 is particularly time-consuming because a high percentage of its signals has multiple fanout branches and, thus, the dominance relation does not substantially reduce the number of candidate pairs substantially. These results are obtained by an event-driven fault simulator [102], which is highly efficient because it only simulates the difference between the faulty circuit and the fault-free circuit for each fault under consideration.

Table 9.2 Results of diagnosing optimized ISCAS85 benchmark circuits injected with **two** errors.

circuit	# signals in C2	# candidate pairs checked			**potential fix pairs**	pessimis. lower bound	CPU-time (seconds)		
		stage1	stage2	total			random simulation	fault simulation	total
C432	175	1035	13	1048	4	2	1	4	5
C499	410	4753	81	4834	12	4	1	12	13
C880	292	3916	32	3948	6	2	1	7	8
C1355	410	4753	36	4789	8	6	1	5	7
C1908	357	4371	361	4732	56	12	1	15	16
C2670	803	2701	135	2836	20	3	2	19	21
C3540	1189	20100	1176	21276	72	8	3	134	137
C5315	1248	34191	5	34196	2	2	3	138	141
C6288	2339	264628	8	264632	4	3	4	3240	3244
C7552	2187	41401	453	41494	9	3	15	1153	1168
Avg.	941	38148	230	38378	19.3	4.5	3	473	476

9.6 Summary

In this chapter, a new approach is introduced to combinational error diagnosis. This algorithm searches for the potential error sources that are most likely to be responsible for the incorrectness of the implementation. Unlike the symbolic approaches, this approach does not rely on the construction of BDDs to search for such signals. Instead, a fault simulation process is used to precisely decide if a signal should be held responsible for the erroneous response of a vector. This formulation allows to exclude most signals from the list of potential error sources efficiently by performing fault simulation with a number of erroneous input vectors. A two-stage algorithm that explores the topological dominance relation between signals is further introduced to speed up the process of diagnosing multiple errors. Compared to other approaches, this method offers three major advantages. First, unlike the symbolic approaches [40,87,88,92,126], the approach is more scalable for larger designs. Secondly, it delivers more accurate results than other simulation-based approaches [73,108,109,123,126] because it is based on a more stringent condition for locating potential error sources. Thirdly, it can be generalized to identify multiple errors.

Chapter 10

Extension to Sequential Error Diagnosis

This chapter extends the fault-simulation-based diagnosis approach to sequential circuits. We first describe the necessary and sufficient condition of a *correctable input sequence* that can be corrected by changing the function of an internal signal. We then introduce a modified sequential fault simulation process to check this condition. This formulation does not rely on any error model and, thus, is suitable for general types of errors. Finally, we discuss the extension to identify multiple errors.

10.1 Introduction

Consider the problem of diagnosing a custom design versus a specification described at a higher level of abstraction, such as the register-transfer (RT) level. Logic synthesis with a low effort can first be performed on the RT-level specification to derive a gate-level representation. Then, a gate-level to gate-level error diagnosis process is followed. It is very likely that a one-to-one flip-flop correspondence between the implementation and a gate-level specification may not exist. Therefore, the combinational diagnosis procedure described in the previous chapter cannot be applied. A sequential error diagnosis approach is needed for this application.

Due to the greater difficulty and complexity, very few papers in the literature have addressed the problem of diagnosing design errors in sequential circuits [48,106,128]. A method, modeling the errors in the state transition table, can only target small controllers [106]. Another approach [48] focuses on small feedback-free circuits, or finite state machines that have a one-to-one state correspondence with their specifications. The approach proposed in [128], extends a combinational backward error tracing heuristic [73,127] for the sequential circuits using the iterative array model. A restricted error hypothesis containing three types of wrong-gate errors is used. This approach may fail when the design error is not modeled in the hypothesis.

10.2 Diagnosing Sequential Circuits

10.2.1 Correctability

The specification can be regarded as a black box with a completely specified input/output behavior. However, for the sake of simplicity without losing generality, it is assumed that both the specification and the erroneous implementation are given as gate-level circuits. If the specification is given at a higher level of abstraction, synthesis with a low effort is first performed to generate a corresponding gate-level specification. The number of flip-flops in the specification and the implementation could be different and no flip-flop correspondence is required in this algorithm.

Definition 10.1 (*erroneous sequence*) A binary input sequence is an erroneous sequence if it causes different output responses at any primary output.

Definition 10.2 (*correctable sequence*) An erroneous input sequence E is said *correctable* by signal f in C_2 if there exists a new function for signal f in terms of the primary inputs and the present state lines of C_2 such that E

becomes a correct sequence for the resulting new circuit (as illustrated in Fig. 10.1).

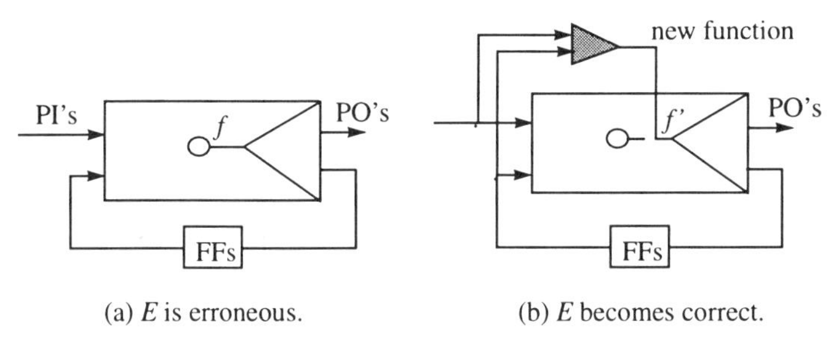

(a) E is erroneous. (b) E becomes correct.

Fig. 10.1 A correctable input sequence E by re-synthesizing a signal.

Based on this definition, we assume that an error source will correct every erroneous input sequence. Initially, every signal is considered a candidate error source. If a signal is found unable to correct *any* erroneous sequence, then it is excluded from the candidate list. It is worth mentioning that, unlike the combinational case, even if a signal can correct every erroneous input sequence, it is not necessarily a single-fix signal for a sequential circuit. For a combinational circuit, every erroneous vector can be fixed *independently*, that is, the requirements to fix all erroneous vectors can be satisfied at the same time. But for a sequential circuit, some conflicts may occur in deriving the fix function for correcting all erroneous sequences simultaneously, even though each of them can be fixed individually.

10.2.2 The Necessary and Sufficient Condition

In the following discussion, we assume that both C_1 and C_2 have known reset states, p_1 and q_1, respectively. The implementation C_2 is represented by the time-frame-expansion model as shown in Fig. 10.2. The number of copies of the combinational portion duplicated in the time-frame-expansion model equals the length of the input sequence under consideration. Consider an erroneous input sequence with 3 input vectors,

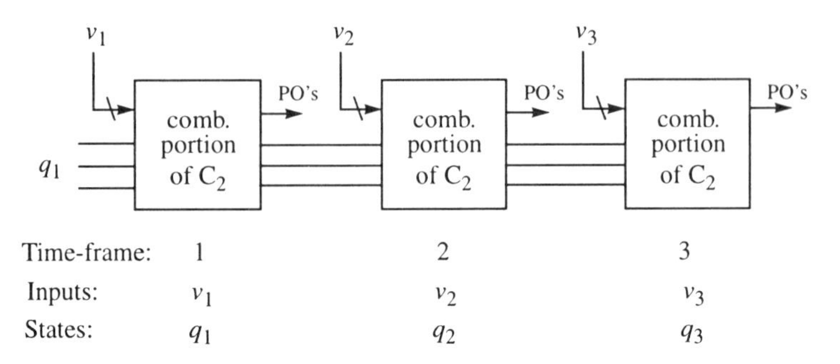

Fig. 10.2 An example of the time-frame expansion model.

$E = \{v_1, v_2, v_3\}$. Suppose E brings C_2 through a sequence of states $\{q_1, q_2, q_3\}$, where q_1 is the initial state of C_2. We call $(v_1 \mid q_1)$ the *pseudo-input vector* for the first time-frame. Similarly, $(v_2 \mid q_2)$ and $(v_3 \mid q_3)$ are the pseudo-input vectors for the second and the third time-frames, respectively. Based on this model, it follows directly that if signal f can correct an erroneous input sequence E, then there exists a new function of f such that every primary output at every time-frame becomes error-free (i.e., every primary output has the same response as its corresponding primary output of C_1 with respect to the input sequence E). We define a term called *injection* before we derive the necessary and sufficient condition of correcting an erroneous input sequence.

Definition 10.3 (*injection*) Given a signal f in C_2, a t *time-frame injection* at f is a set of value assignments to the signal f for the first t time-frames in the array model. For example, $J = \{f^{(1)} = 0, f^{(2)} = 0, f^{(3)} = 0\}$ represents a 3 time-frame injection that injects value '0' at f for all 3 time-frames, where the superscript denotes the index of a time-frame.

An injection defines a new circuit. The output responses of the resulting new circuit are computed by treating the injected signals as an independent pseudo primary input lines, taking the injected constants as the values. Since we can inject either '0' or '1' to a signal at each time-frame, there are

2^t different combinations for a t time-frame injection. The complexity grows exponentially with the number of time-frames.

Proposition 10.1 Let E be an erroneous input sequence with t input vectors, $E = \{v_1, v_2, ..., v_t\}$. A signal f in C_2 can correct E *only if* there exists a t time-frame injection at f such that E becomes a correct input sequence for the resulting new circuit. Such an injection is called a *cure injection* at f for E.

Explanation: This proposition states that if there exists no cure injection at f, then there exists *no new function* for f to correct the erroneous input sequence E, but not vice versa. In other words, to correct an erroneous input sequence, it is necessary to find a cure injection. However, a cure injection is not sufficient to assure that the input sequence is indeed correctable. This is due to the fact that, for every fix function at f, there always exists a cure injection. Conversely, not every cure injection can be realized by a function.

Given a fix function at f that can correct an erroneous input sequence E, the corresponding cure injection can be derived as follows: Let the responses at f with respect to E in the resulting new circuit is $\{\alpha_1, \alpha_2, ..., \alpha_t\}$, where α_i, $i \le i \le t$, is a binary value. Then $J = \{f^{(1)} = \alpha_1, f^{(2)} = \alpha_2, ..., f^{(t)} = \alpha_t\}$ is a cure injection. On the other hand, there exists some injection that is not realizable. Fig. 10.3 shows an example. The pseudo-input vectors for the first and second time-frames are the same: (0 | 000). But the injected values at f for these two time frames are different ('0' and '1', respectively). A function realizing this injection needs to map the same pseudo-input vector, (0 | 000), to '0' and '1' at the same time, which is impossible. It follows that there exists no new function for f in terms of primary inputs and present state lines to realize this injection. Whether or not an injection is realizable can be easily checked by simulating the input vector for the resulting circuit with the injection. After collecting the sequence of states encountered in the resulting circuit, the pseudo-input vector for each time-frame and the injected value can then be derived. If no conflict exists, then the injection is realizable.

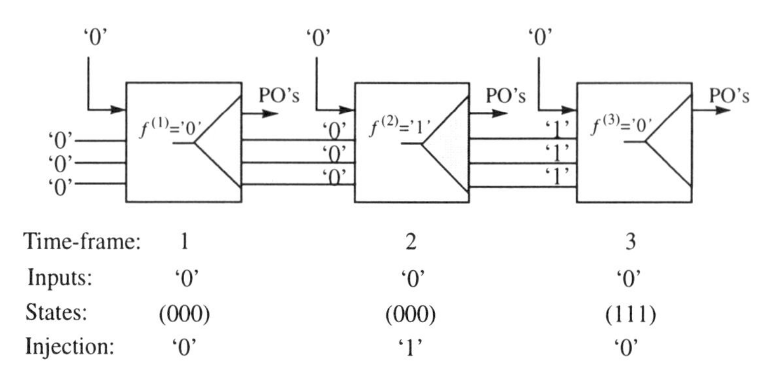

Fig. 10.3 Unrealizable injection.

Proposition 10.2 (the necessary and sufficient condition) Let E be an erroneous input sequence with t input vectors, $E = \{v_1, v_2, ..., v_t\}$. A signal f in C_2 can correct E if and only if there exists a *realizable t time-frame cure injection*.

10.2.3 Correctability Check via Fault Simulation

Based on the above proposition, we can determine if an erroneous input sequence is correctable by a signal in two steps: (1) check if there exists a cure injection, and (2) check if it is realizable. Given an injection, determining whether it is a cure injection can be done via a modified sequential fault simulation process. Traditionally, sequential fault simulation assumes that the target signal is stuck at the same binary value for every time-frame. In this application, a signal is allowed to be stuck at different binary values at different time-frames for fault simulation to account for the fact that an injection may have different values at different time-frames.

In the worst case, we need to enumerate every possible 2^t injections in the worst case to find a cure injection, or to conclude that there does not exist one. However, like most branch-and-bound procedures, some bounding criterion can be used to narrow down the search space. In this application, the search space can be represented as a binary injection tree shown in Fig. 10.4. The meaning of this tree is explained as follows: (1) A node in the tree corresponds to a resulting circuit with a partial injection (i.e., t' time-frame injection where $t' < t$). The root node (level 0) corresponds to the original implementation C_2. (2) The level of a node corresponds to the current time-frame being considered for value injection. (3) The upper (lower) branch from a node represents injecting value '0' ('1') to the signal under consideration in the current time-frame. (4) A path from the root node to a leaf node represents a complete t time-frame injection.

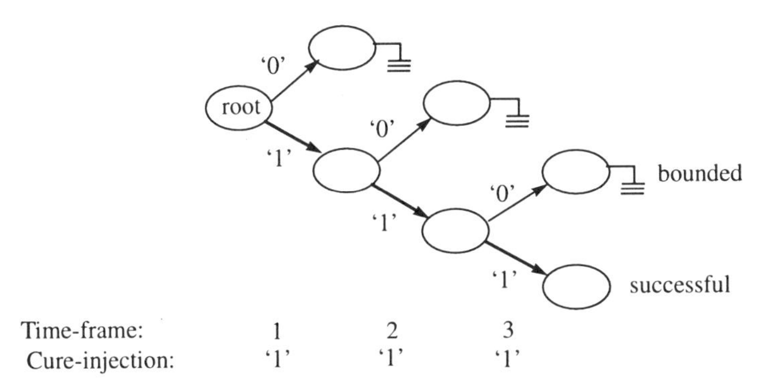

Fig. 10.4 Injection tree.

Based on this injection tree, it follows that if a node's corresponding partial t' time-frame injection cannot produce correct responses for the first t' time-frames, then the sub-tree of this node need not be further explored. This simple bounding criterion is useful in speeding up the search. Once a cure injection is found, the pseudo-input vectors of the resulting circuit with respect to that injection can be derived. The realizability check is then followed to determine if the erroneous input sequence is indeed correctable by a realizable function at the target signal.

10.2.4 Generalization for Multiple Errors

For diagnosing circuits with multiple errors, this algorithm searches for multiple signals that can jointly correct every pre-generated erroneous input sequence. Fig. 10.5 shows an example of a 3 time-frame injection defined over a set of signals {*f*, g}. Similar to the case of single-error diagnosis, if

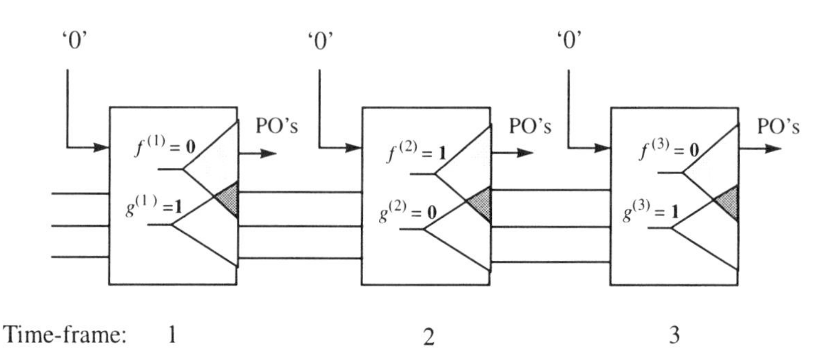

Fig. 10.5 A 3-time-frame injection defined over a set of signals {*f*, *g*}.

this injection is a realizable cure injection for the applied erroneous sequence *E* = {0, 0, 0), then, *E*, is correctable by this set of signals. Again, this condition can be checked primarily via a modified fault simulation process. Given a set of *k* signals, σ, and an erroneous input sequence with *t* input vectors, *E*, the worst-case complexity of deciding if *E* can correct σ is proportional to the number of possible injections. There are 2^k possible value combinations for each time-frame and, thus, the total number of possible *t* time-frame injections defined over σ is $2^{(k \cdot t)}$.

In a previous chapter, we introduce a two-stage multiple-error diagnosis algorithm for combinational circuits. The extension of that algorithm for sequential circuits is straightforward.

10.3 Experimental Results

This extension has been implemented in *ErrorTracer*, and has been tested on ISCAS-89 sequential benchmark circuits. Each circuit was optimized by script *script.rugged* to obtain the implementation. Then it was decomposed into AND/OR gates using SIS command "*tech_decomp -a 5 -*

o 5". The gate type errors are injected randomly using a program [73] to generate erroneous implementations. Note that this algorithm does not require the knowledge of the number of flip-flops or the structure of the specification. Only the input/output functional behavior of the specification is needed.

10.3.1 Results of Diagnosing Single-error Circuits

Table 10.1 shows the results of single-error diagnosis. In the pre-processing stage, random simulation is performed to collect erroneous

Table 10.1 Results of diagnosing optimized ISCAS-89 benchmark circuits injected with **one** error.

circuit	# signals in C2	E-length (min.)	# potential fix signals	pessimistic lower bound	CPU-time (seconds)		
					random simulation	fault simulation	total
s298	65	10	3	2	36	13	49
s344	102	4	4	2	1	12	13
s349	101	3	2	1	1	10	11
s386	60	1	1	1	1	2	3
s510	146	5	7	4	1	12	13
s641	109	1	2	2	1	1	2
s713	112	1	2	2	1	5	6
s820	171	5	7	3	9	12	21
s832	174	2	2	2	1	8	9
s1196	327	1	6	5	1	13	14
s1238	348	1	1	1	1	8	9
s1488	387	7	18	2	205	40	245
s1494	375	2	11	2	2	27	29
s1423	422	10	42	2	209	1020	1229
s5378	920	2	3	2	13	53	66
s9234	1075	6	6	2	38	1050	1089
s13207	1325	2	4	2	11	205	216
Avg	631	3.7	7.1	2.2	31	146	178

Intractable circuits:

(1) s208, s400, s444: due to long erroneous sequences.

(2) s15850, s35932, s38417, s38584: due to difficulty of generating erroneous sequences.

input sequences. The maximum limit on the length of the sequences is set to 10. The random simulation terminates when either 32 erroneous input sequences have been collected, or 32,000 sequences have been simulated.

The meanings of some columns in Table 10.1 are as follows. Column "E-length (min)" is the minimal length of the erroneous input sequences found in the pre-processing step. Column "*# potential fix signals*" is the number of the potential single-fix signals delivered by ErrorTracer. On average, the number of the potential single-fix signals produced by ErrorTracer is 7.1 for the listed ISCAS-89 benchmark circuits. The curves of the number of the potential single-fix signals versus the number of simulated erroneous input sequences for s1196 and s5378 are shown in Fig. 10.6.

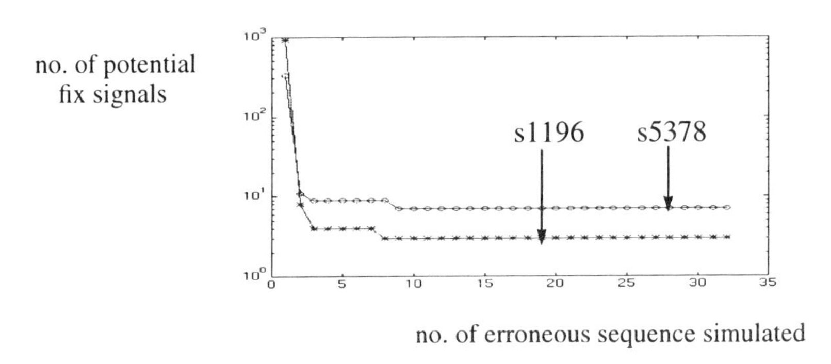

Fig. 10.6 The number of potential single-fix signals versus the number of fault simulated erroneous input sequences.

We observe that only a small number of erroneous input sequences is enough to filter out most false single-fix candidates. The program fails on seven circuits: s208, s400, s444, s15850, s35932, s38417, and s38584. The reasons will be discussed later. Column "*Lower bound*" is a *pessimistic* lower bound on the total number of single-fix signals. For ISCAS-89 benchmark circuits, the average is 2.2. Column "*Time*" is the CPU-time on a 150 MHz Sparc20 workstation. The total time consists of the random simulation time and the fault simulation time. These results are obtained by using a modified differential fault simulation algorithm [36].

10.3.2 Results of Diagnosing Double-Error Circuits

Table 10.2 shows the results of diagnosing the implementations injected with two random errors. ErrorTracer first searches for the single-fix signals. If there exists no such signals, then the search for signal pairs that can jointly correct the implementation is activated. Four out of the 15 circuits in Table 10.2 are classified as single-signal correctable. Note that these circuits may or may not be indeed single-signal correctable. It is possible that none of the candidate single-fix signals is a real single-fix signal. On the other hand, the other 11 are proven *not* to be single-signal correctable by *ErrorTracer* (i.e., the number of potential single-fix signals is 0). For

Table 10.2 Results of diagnosing optimized ISCAS-89 benchmark circuits injected with **two** error.

circuit	# signals in C2	E-length (min.)	#potential single-fix	#potential double-fix	pessimi. lower bound	time (seconds)		
						random simulation	fault simulation	total
s298	65	2	0	36	4	1	212	213
s344	102	3	0	16	6	1	611	612
s349	101	2	0	16	6	1	60	61
s386	60	1	3	-	-	2	42	44
s510	146	1	0	5	2	1	100	101
s641	109	1	0	15	6	1	60	61
s713	112	1	0	6	4	1	26	27
s820	171	2	1	-	-	1	10	11
s832	174	2	3	-	-	4	6	10
s1196	327	2	0	10	3	1	86	87
s1238	348	3	0	10	4	1	156	157
s1488	387	1	0	2	1	1	268	269
s1494	375	2	0	18	1	2	312	314
s1423	422	3	5	-	-	7	109	116
s5378	920	1	0	14	4	10	3873	3883

these circuits, the number of the potential double-fix pairs is reported. The CPU time is longer than the case of diagnosing single errors due to the rapid growth in the number of candidate signal pairs and the number of possible injections that need to be checked for correctability.

10.3.3 Challenges

Several issues need to be further addressed in the future:

(1) *Erroneous input sequence generation.* For combinational circuits, random simulation [72] or automatic test pattern generation (ATPG) techniques [1,56] have provided satisfactory solutions to the generation of erroneous vectors even for fairly large circuits. However, these techniques are not adequate for some large or highly sequential designs. Random simulation could not find any erroneous input sequence for the single-error implementations s15850, s35932, s38417, and s38584 in this experiments. For these designs, if manually crafted functional sequences are available for simulation-based design validation, they can be used to identify erroneous sequences. Another possible solution to this problem is to first explore the flip-flop correspondence or internal structural similarity between the specification and implementation. If a large number of matching flip-flop pairs exists, then the sequentiality of the circuit can be reduced by treating the inputs (outputs) of matching flip-flops as pseudo-primary outputs (inputs). In this way, the difficulty of generating erroneous input sequences can be substantially reduced.

(2) *High complexity for processing long erroneous input sequences.* If errors occur in a highly sequential module (e.g., a counter) and cannot be detected by any input sequence with a reasonable length (e.g., 20 vectors), then this approach may become too time-consuming. For s208, s400 and s444, this approach fails due to this reason.

(3) *Difficulty for diagnosing circuits with errors of high cardinality.* In practice, the complexity of diagnosing circuits with more than two errors is prohibitively high. Some heuristics are needed to provide the designer an estimate of each signal's error probability.

10.4 Summary

In this chapter, we have discussed a fault-simulation-based approach for diagnosing design errors in a sequential circuit. This approach allows the implementation to have a different number of flip-flops or a different state encoding from the specification. This capability is particularly useful when the specification is given at a higher level of abstraction such as the register-transfer level. This approach uses a modified fault simulation procedure to determine if a signal can be held responsible for a particular erroneous input sequence. Based on this formulation, this algorithm excludes most signals from being error sources by simulating a number of erroneous input sequences. Preliminary experimental results on diagnosing the ISCAS-89 benchmark circuits with one and two random errors show that this could be an important step towards sequential error diagnosis while several issues still need to be addressed for large and highly sequential designs.

Chapter 11

Incremental Logic Rectification

Both engineering change and error correction can be formulated as a logic rectification problem. This problem takes two gate-level netlists, which are functionally inequivalent, as inputs. One is considered as the specification and the other the implementation. The objective is to find a transformation to apply on the implementation such that the resulting new implementation is functionally equivalent to the specification, while the logic in the old implementation is reused as much as possible to realize the a new implementation.

We describe a hybrid approach for logic rectification. Four major features of this approach will be explained in this chapter. (1) The local BDD-based verification techniques [60] are first applied to trim down the potential error region, so that the error correction process can focus on a smaller region. (2) Using the symbolic analysis, a necessary and sufficient condition of *partial correction*, which is an operation to reduce the size of the input vector set differentiating the specification and implementation, is derived. Based on this condition, the implementation is rectified by a sequence of partial corrections. This rectification process gradually and incrementally transforms the implementation into a circuit equivalent to the specification. (3) A necessary and sufficient condition of general single-gate correction is derived. This condition is particularly useful when there are gate-type errors in the implementation. (4) The re-synthesis based approach is integrated with the structural approach to handle larger circuits for which the structural correspondence between the specification and the implementation is available.

11.1 Preliminaries

11.1.1 Definitions

As in previous chapters, the primary inputs for both circuits are denoted as $\{x_1, x_2, ..., x_n\}$. The primary outputs of the specification and implementation are denoted as $\{S_1, S_2, ..., S_m\}$ and $\{I_1, I_2, ..., I_m\}$, respectively. We assume that there is a one-to-one correspondence between the primary outputs of the specification and the implementation. The definitions of signal pair (Def. 3.1), and the erroneous vector (Def. 8.2) as given in earlier chapters will be used in the following discussion.

Definition 11.1 (*difference set*) The difference set for i-th primary output pair, denoted as $DIFF_i$, is the set of input vectors that differentiate the i-th primary output pair (S_i, I_i), where $0 \leq i \leq m$.

11.1.2 Single Signal Correctable Circuit

For simplicity, we first assume that both C_1 and C_2 are single output functions to explain a necessary and sufficient condition for single signal correctable circuits. This explanation takes a different viewpoint from the one in Chapter 8. Later, we will generalize the condition for circuits with multiple outputs. As illustrated in Fig. 11.1(a), if there exists an internal signal f in C_2 such that the sensitization set (Def. 9.2) $SEN_1(f)$ covers the entire difference set $DIFF_1$ then C_2 can be fixed by changing the function of f. On the other hand, if the sensitization set of a signal h does not cover $DIFF_1$ (as illustrated in Fig. 11.1(b)), then h is not a single-fix signal.

The new function for a single-fix signal f in order to fix the entire C_2 should satisfy two conditions: (1) it should *disagree* with the original

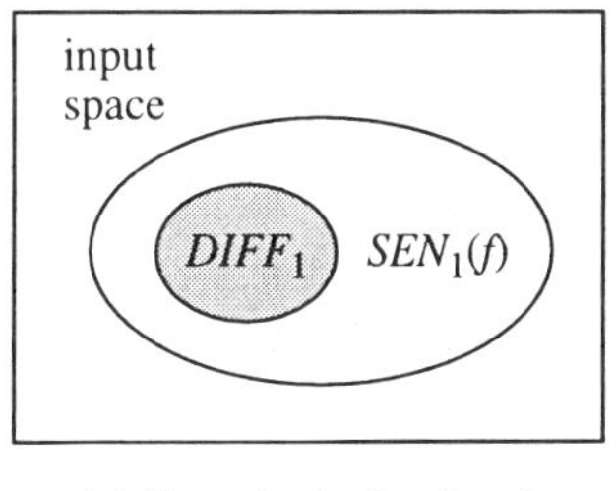

(a) f is a single-fix signal.

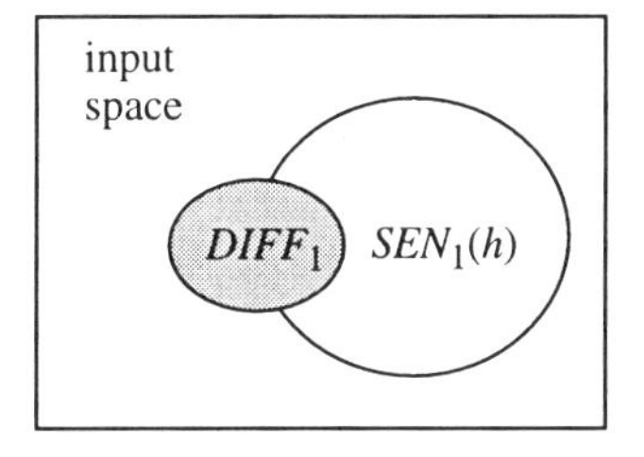

(a) h is a *not* a single-fix signal.

Fig. 11.1 A necessary and sufficient condition of single-fix signal for single-output circuits.

function of f for the input vectors in the difference set, $DIFF_1$, so that the wrong response of each erroneous vector is corrected, and (2) it should *agree* with the old function of f for the input vectors in the sensitization set but not in the difference set, $(SEN_1 - DIFF_1)$, so that no new erroneous vector is created. Let the on-set (off-set) of the original function of f be f^{on} (f^{off}). Then, we can represent the on-set and off-set of the incompletely specified new function for f as follows:

$$f^{new\text{-}on} = (f^{off} \cap DIFF_1) \cup (f^{on} \cap (SEN_1 - DIFF_1))$$

$$f^{new\text{-}off} = (f^{on} \cap DIFF_1) \cup (f^{off} \cap (SEN_1 - DIFF_1))$$

For circuits with multiple outputs, each primary output will impose constraints on the on-set and the off-set. These constraints should be combined together to derive the final on-set and off-set for the new function.

Proposition 11.1 Let i be the index of a primary output $(0 \le i \le m)$, ε be an internal signal of the implementation. The set of input vectors $\varepsilon^{new\text{-}on}$ and

$\varepsilon^{new\text{-}off}$ are defined as follows.

$$\varepsilon^{new-ON} = \bigcup_i ((\varepsilon^{OFF} \cap DIFF_i) \cup (\varepsilon^{ON} \cap (SEN_i - DIFF_i)))$$

$$\varepsilon^{new-OFF} = \bigcup_i ((\varepsilon^{ON} \cap DIFF_i) \cup (\varepsilon^{OFF} \cap (SEN_i - DIFF_i)))$$

If the intersection of $\varepsilon^{\text{new-on}}$ and $\varepsilon^{\text{new-off}}$ is empty, then ε is a single-fix signal.

11.2 Incremental Logic Rectification

In this section we first discuss the procedure of narrowing down the error region by using the verification techniques. Then, we describe a structural approach for incremental logic rectification. After that, we introduce a symbolic approach based on the concept of partial correction. Also, we formulate the single gate correction condition to improve the quality. Finally, we propose a hybrid approach that combines the structural approach and the symbolic approach.

11.2.1 Error Region Pruning

The incremental verification technique used in AQUILA can be applied to narrow down the error region before logic rectification. Consider a simple case where the incorrectness of the implementation is due to two errors as illustrated in Fig. 11.2. We have observed that, even after an implementation is intensively optimized, certain degree of structural similarity still exists for most designs. Due to the error effects, we will not be able to find equivalent signal pairs in the transitive fanout cones of the error signals. However, many equivalent signal pairs outside the transitive fanout cones of the error signals still exist. Through the functional equivalence checking techniques described in Chapter 3, the equivalent signal pairs are identified and merged. Once that is completed, the joint network is traversed from primary outputs toward primary inputs to find a cutset closest to the primary outputs consisting of only the previously identified

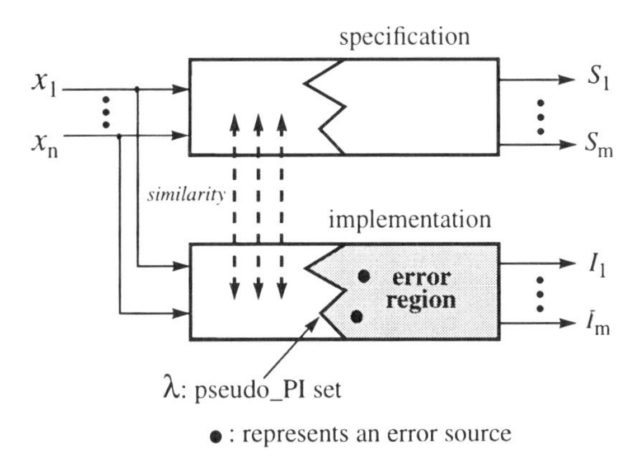

Fig. 11.2 Error region pruning by identifying equivalent signal pairs.

merge points or primary inputs. This cutset is called a *pseudo_PI set*, denoted as λ. In the subsequent symbolic rectification process, only the region to the right of λ is considered as the potential error region (as shown in Fig. 11.2). This error-region pruning technique can also be viewed as an automatic recycling process of the sub-circuit in the implementation that is not affected by the errors. Fig. 11.3 shows an example of the joint network before and after applying this technique. After the error region pruning, only signal c_2 and I_1 are considered as potential error signals.

11.2.2 Symbolic Partial Correction

Property 11.1 *The correctable set* for an internal signal f with respect to a primary output I_i in the implementation, denoted as $R_i(f)$, is the intersection of its corresponding sensitization set $SEN_i(f)$, and difference set $DIFF_i$.

Intuitively, the correctable set $R_i(f)$ represents a *maximal set of erroneous vectors* with respect to the i-th primary output that can be corrected by changing the function of signal f.

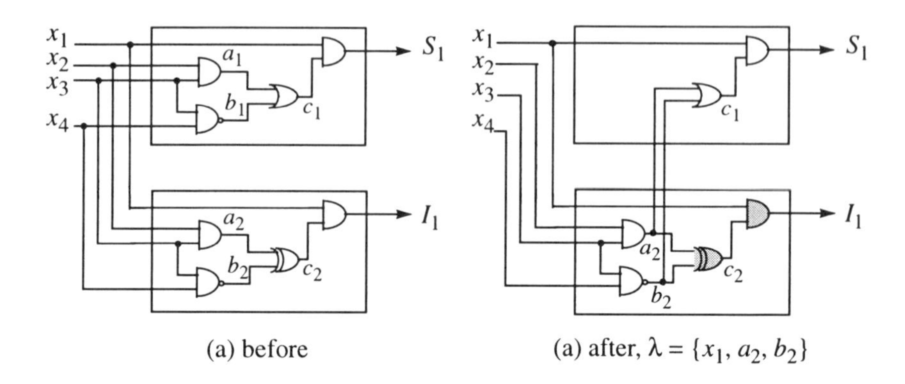

Fig. 11.3 The joint network before and after error-region pruning. (The error region is indicated in shadow)

Definition 11.2 (*partial correction*) The re-synthesis of an internal signal f in the implementation is called a partial correction if two conditions are satisfied: (1) no erroneous vector is newly created for any primary output, and (2) the difference set for at least one primary output is reduced. The signal f is called a *partial-fix signal.*

Based on this definition, the rectification process can be viewed as a sequence of partial corrections as shown Fig. 11.4. During this process, the

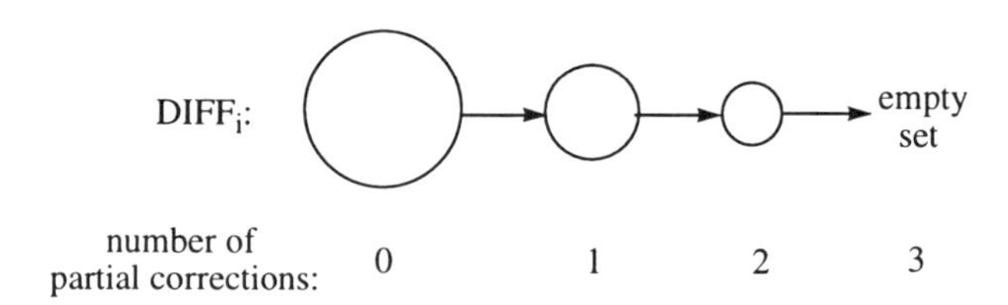

Fig. 11.4 Incremental logic rectification process through a sequence of partial corrections.

difference set with respect to each primary output monotonically shrinks until it becomes empty. Each partial correction consists of finding a partial-fix signal and synthesizing the new fix function to replace the old function. Note that partial correction is a weak criterion and, thus, there may be too many choices in selecting a partial-fix signal. Most of these partial-fix signals do not deliver high quality fixes. Therefore, we define strong partial correction.

Definition 11.3 (*strong partial correction*) The re-synthesis of an internal signal f in the implementation is called a strong partial correction if two conditions are satisfied: (1) no erroneous vector is newly created, and (2) the correctable set of f with respect to every primary output is corrected.

Fig. 11.5 illustrates the construction process of the new function for a strong partial-fix signal f. For each primary output, the input space (in terms of the *pseudo_PI set*) is partitioned into three regions: (I) the *don't care region*, (II) the *fix region*, and (III) the *don't touch region*. These three regions are determined by the sensitization set of f w.r.t. the primary output under consideration, I_i, and the difference set for the i-th primary output pair.

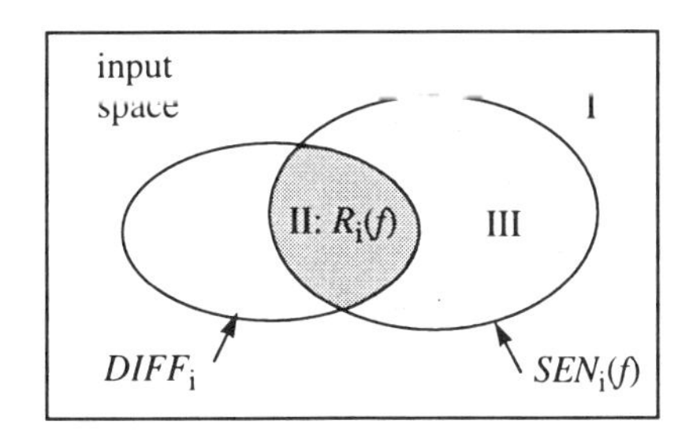

Region I: $\sim(SEN_i(f))$
Region II: $R_i(f)$
Region III: $SEN_i(f) - R_i(f)$

Fig. 11.5 Computing the new function for a strong partial correction.

The don't care region consists of the input vectors that are not in the sensitization set $SEN_i(f)$. Changing the function of f with respect to any input vector in this region does not affect the i-th primary output response. Hence, this is also the don't care set for the new function.

The second region, namely the fix region, is the set of erroneous vectors to be fixed by the new function. The fix region is simply the correctable set $R_i(f)$. In order to fix the i-th primary output response with respect to any input vector, the new function should be complementary to the original function in this region.

The third region, called the don't touch region, is the set of input vectors for which the new function should be exactly the same as the original function. Otherwise, new erroneous vectors will be created. Any input vector in the sensitization set, but not in the difference set is a don't touch vector. This is because if an input vector v is in the don't touch region, it means that the responses of the i-th primary output pair (S_i, I_i) are identical with respect to that vector. Since this vector is also in the sensitization set, changing the function of f with respect to v also changes the response of I_i. Therefore, the responses at S_i and I_i will become different. The constraints of strong partial correction for all primary outputs in the implementation should be combined together.

Proposition 11.2 (*necessary and sufficient condition for strong partial correction*) Let i be the index of a primary output, $0 \le i \le m$, and ε be an internal signal in the implementation. Let $\varepsilon^{new\text{-}on}$ and $\varepsilon^{new\text{-}off}$ be two sets of input vectors defined as:

$$\varepsilon^{new-on} = \bigcup_i ((\varepsilon^{off} \cap R_i) \cup (\varepsilon^{on} \cap (SEN_i - R_i)))$$

$$\varepsilon^{new-off} = \bigcup_i ((\varepsilon^{on} \cap R_i) \cup (\varepsilon^{off} \cap (SEN_i - R_i)))$$

where $\varepsilon^{\mathrm{on}}$ and $\varepsilon^{\mathrm{off}}$ are the on-set and the off-set of the original function at ε. If the intersection of $\varepsilon^{\mathrm{new\text{-}on}}$ and $\varepsilon^{\mathrm{new\text{-}off}}$ is empty, then ε is a strong partial-fix signal, and the new function is an incompletely specified function whose on-set and off-set are defined by $\varepsilon^{new\text{-}on}$ and $\varepsilon^{new\text{-}off}$, respectively.

11.2.3 Single-Gate Correction Criterion

Once a strong partial-fix signal is found, the original function is replaced by the new function. The new function is represented in terms of the pseudo_PI set (derived in the error-region pruning stage). In order to further improve the recycling rate, a new formulation is derived to check if the new function can be realized by only changing the function of one logic gate. As shown in Fig. 11.6, let ε be the strong partial-fix signal, and G be its fanin

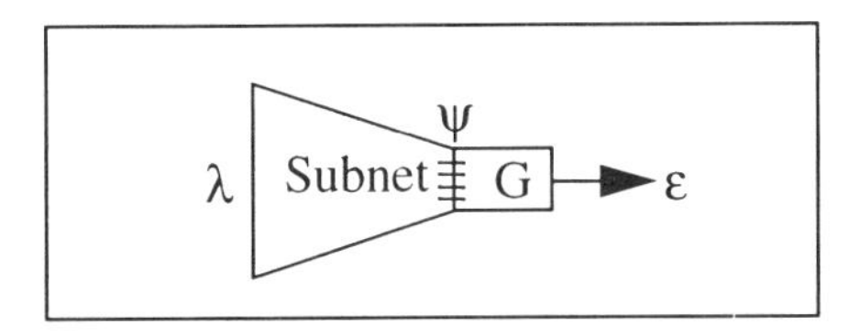

λ: the pseudo_PI set
ψ: the set of immediate fanins of G.

Fig. 11.6 Illustration of the single gate correction condition.

gate. The immediate fanin set of G is denoted as ψ. The *subnet* is defined as the sub-network taking the pseudo_PI set as the inputs and ψ as the outputs as shown in Fig. 11.6. The objective here is to recycle the subnet to realize the desired function ε^{new}. In other words, we need to find a new logic gate G^{new} such that the composite function (*Subnet*)(G^{new}) realizes the desired function ε^{new}. The first step in finding the new gate function G^{new} is to convert the support of the on-set and off-set of ε^{new} from λ to ψ via image computation associated with the *subnet*. Let

$$G^{new-ON} = Image(\varepsilon^{new-ON})$$

$$G^{new-OFF} = Image(\varepsilon^{new-OFF})$$

Note that $\varepsilon^{new\text{-}on}$ and $\varepsilon^{new\text{-}off}$ are two sets of vectors defined over λ, while their images associated with the network *subnet*, $G^{new\text{-}on}$ and $G^{new\text{-}off}$, are sets of vectors defined over ψ. The single gate correction condition can be stated as follows.

Proposition 11.3 (*single gate correction condition*) Based on the notation of Fig. 11.6, if $G^{new\text{-}on}$ and $G^{new\text{-}off}$ are not disjoint, then there exists no single logic gate G' such that (*Subnet*)(G') realizes the desired function ε^{new}. On the other hand, if $G^{new\text{-}on}$ and $G^{new\text{-}off}$ are disjoint, then $G^{new\text{-}on}$ and $G^{new\text{-}off}$ are the new on-set and the off-set such that the composite network (*Subnet*)(G^{new}) realizes the desired function ε^{new}.

Proof: *Subnet* is a mapping from λ to ψ, and G is a function from ψ to {0,1} as shown in Fig. 11.7.

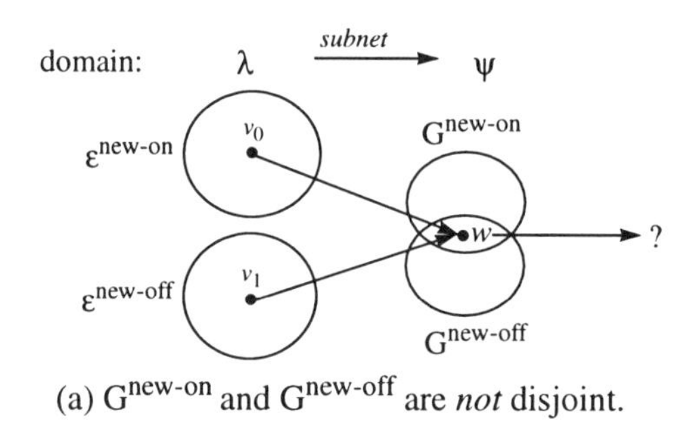

(a) $G^{new\text{-}on}$ and $G^{new\text{-}off}$ are *not* disjoint.

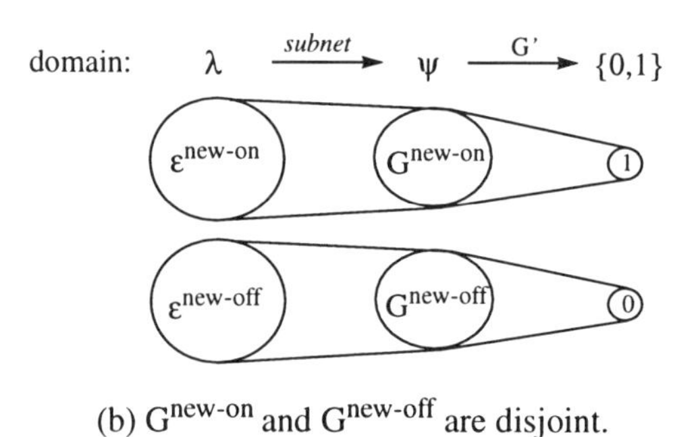

(b) $G^{new\text{-}on}$ and $G^{new\text{-}off}$ are disjoint.

Fig. 11.7 Proving the single gate correction condition.

(1) Suppose $G^{new\text{-}on}$ and $G^{new\text{-}off}$ are not disjoint. Then there exist two vectors, v_0 and v_1, in terms of λ such that $v_0 \in \varepsilon^{new\text{-}off}$, $v_1 \in \varepsilon^{new\text{-}on}$ and $subnet(v_0) = subnet(v_1) = w$, where w is a vector in terms of ψ. As shown in Fig. 11.7(a), it is impossible to find a new function G' from ψ to $\{0,1\}$ such that the composite function $(subnet)(G')$: $\lambda \rightarrow \{0,1\}$ maps v_1 and v_0 to 1 and 0, respectively. Hence, there does not exist a new logic gate G' to satisfy the single gate correction condition.

(2) If $G^{new\text{-}on}$ and $G^{new\text{-}off}$ are disjoint, then composite function $(subnet)(G^{new})$: $\lambda \rightarrow \{0,1\}$ realizes the desired function as shown in Fig. 11.7(b). (Q.E.D.)

11.2.4 The Algorithm

Fig. 11.8 shows the overall flow. This algorithm iterates until every primary output pair of the specification and the implementation is equivalent. Each iteration has two phases: (1) error-region pruning using the verification techniques, and (2) symbolic partial correction. In the first phase, we first use the local-BDD based incremental verification techniques of AQUILA (Chapter 5) is used to identify and merge equivalent signal pairs. A pseudo_PI set consisting of a set of identified equivalence points closest to primary outputs is selected. Only signals in the transitive fanout of this pseudo_PI set are considered as potential error signals. In the second phase, the potential error signals are examined one-by-one in a *fanin-first order* until a strong partial-fix signal is found. In the worst case, a primary output of C_2 may be selected as the next fix signal. To check the strong partial correction criterion, the BDDs of the sensitization set for the target signal with respect to each primary output, and the difference set for each primary output pair in terms of the pseudo_PI set are constructed. Once a partial-fix signal is found, its fix function based on Proposition 11.2 is derived. Since the desired new function at a strong partial-fix signal is usually an incompletely specified function, it is represented by two BDDs characterizing the on-set and the off-set, respectively. These two BDDs are then converted into a sub-network implementing the desired function in

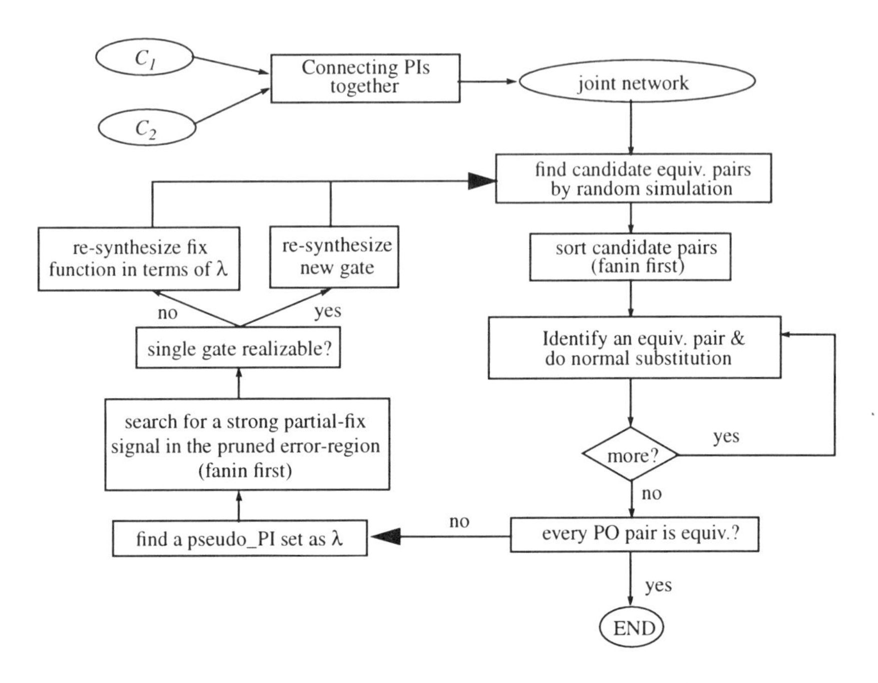

Fig. 11.8 The verall flow of the symbolic logic rectification.

terms of the pseudo_PI set. Finally, the old signal is replaced by this newly generated sub-network. If applicable, the single gate correction condition is used for the new function to achieve a higher recycling rate.

11.3 A Divide and Conquer Heuristic

The above algorithm may not be applicable to large circuits because it requires the construction of BDDs for the characteristic functions of the difference set and the sensitization set. Although these BDDs are built in terms of the pseudo_PI set, they cannot be constructed for large circuits when errors occur close to primary inputs. Hence, the structural information is incorporated to further extend its capability. The structural information here refers to signal pairs between the specification and the implementation. If some pairs are specified by the user or are assumed through identical naming, then the problem becomes easier. The logic rectification problem can be solved in a divide-and-conquer manner. At each iteration, the algorithm selects a closest cutset consisting only of corresponding pairs in the fanout of the pseudo_PI set as shown in Fig. 11.9. These pairs are then treated as the pseudo primary output pairs to perform the subsequent symbolic partial correction.

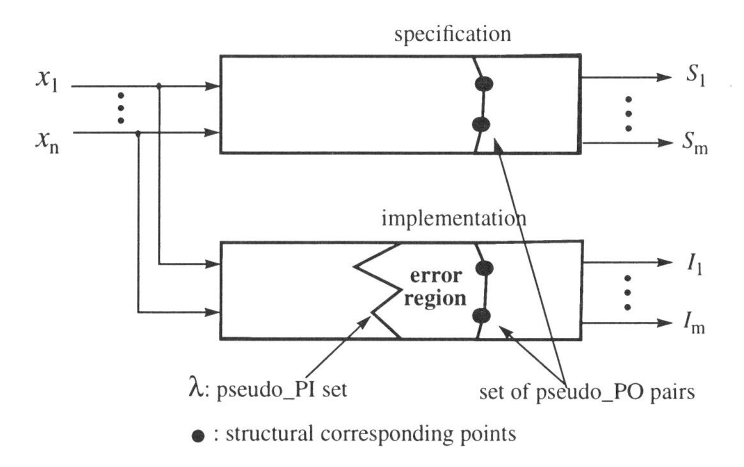

Fig. 11.9 Incorporation of the structural correspondence information to narrow down the region for the symbolic partial correction.

11.3.1 Pure Structural Approach Based on Back-Substitution

In this subsection, we describe a structural approach for large circuits. It relies on back-substitution to rectify the circuits gradually [14,60]. In the following, signals a_1 and a_2 are referred to as *key signals* if (a_1, a_2) is a structural corresponding pair (specified by the user or provided through naming information). The overall procedure shown in Fig. 11.10 consists of three major stages: structural correspondence extraction, error diagnosis, and correction. Similar to the symbolic algorithm described in Fig. 11.8, multiple iterations of error diagnosis and correction may be required to completely rectify the circuit. At each iteration, the algorithm first identifies and merges equivalent pairs in stages. Then, a heuristic is applied to predict a potential error source in the implementation. In this algorithm, only key

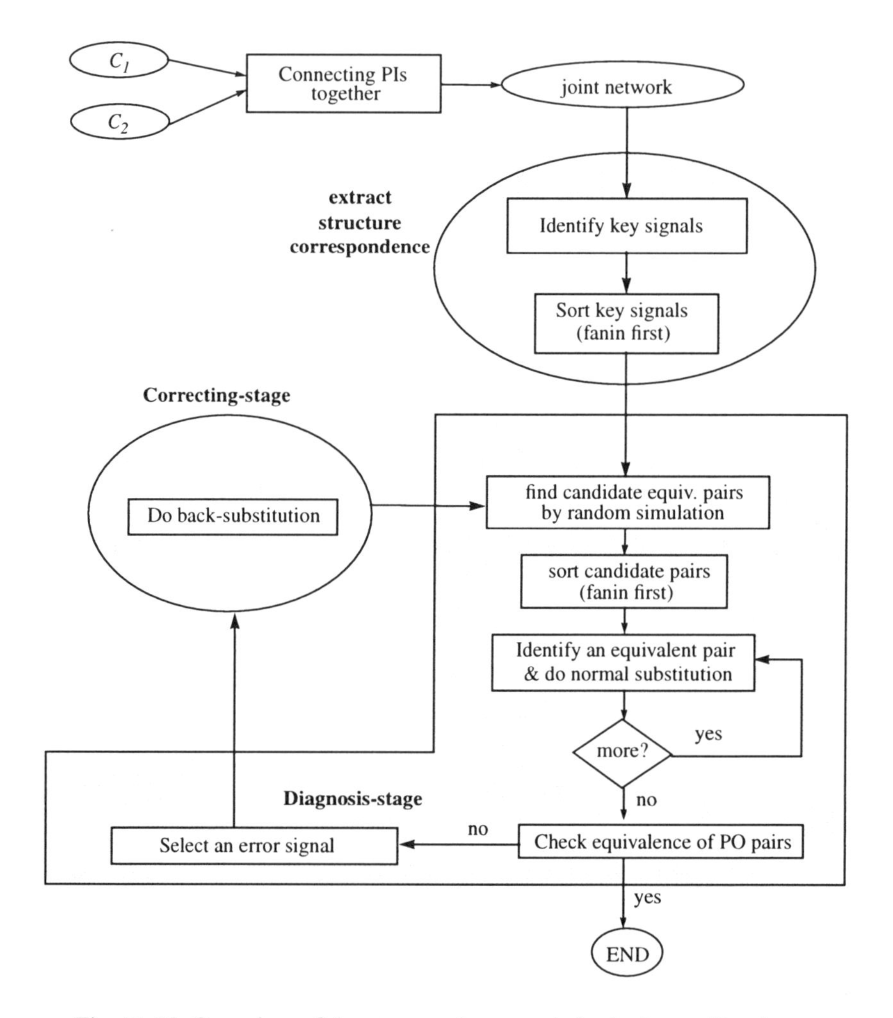

Fig. 11.10 Overview of the structural approach for logic rectification.

signals in the implementation could be considered as potential error sources. Once an error source is selected, signal ε_2 is replaced by ε_1, where ε_1 is the structural corresponding signal of ε_2. By a sequence of such back-substitutions, the primary output pairs become equivalent eventually. In the worst case, the back-substitution is performed on every primary output pair.

In the following, we describe the details of the heuristic to select a potential error source. The heuristic first searches for the closest cutset starting from primary outputs toward primary inputs such that every signal α in this cutset satisfies two conditions: (1) α is a key signal and (2) α is an equivalence point (i.e., has been previously identified as equivalent and merged with its corresponding signal). Finding such a cutset is helpful in locating the error sources. Fig. 11.11 shows an example. The key signal

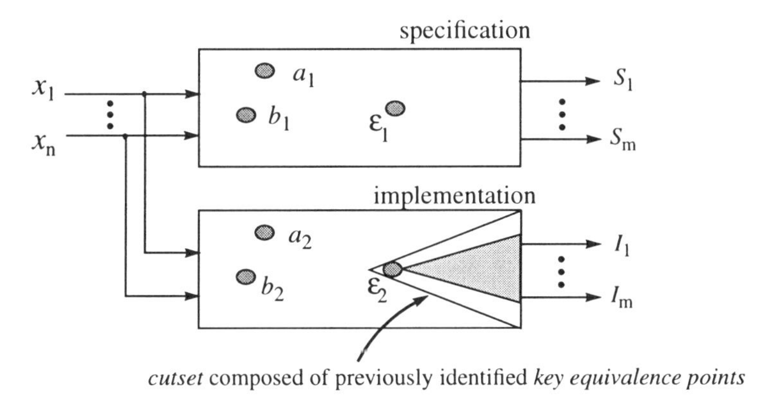

Fig. 11.11 Selecting a potential error source for back-substitution.

pairs that were regarded as inequivalent during the equivalent signal identification process are indicated in shadow. These includes (a_1, a_2), (b_1, b_2), $(\varepsilon_1, \varepsilon_2)$, and a number of key signal pairs in the fanout cone of $(\varepsilon_1, \varepsilon_2)$. Suppose a cutset that envelops the signal pair $(\varepsilon_1, \varepsilon_2)$ and its fanout cone is found by this heuristic as illustrated in the figure. This indicates that the inequivalences of signal pairs (a_1, a_2) and (b_1, b_2) do not cause their output cones to become erroneous and, thus, are not observable at the outputs. In other words, the functional mis-matches at (a_1, a_2) and (b_1, b_2) are not responsible for the incorrect output function of the implementation.

On the other hand, the output cone of signal ε_2 is affected because of the functional mis-match at signal pair $(\varepsilon_1, \varepsilon_2)$, which results in the incorrectness of primary outputs. Therefore, signal ε_2 is regarded as a potential error source. To find such an error source, we examine each key signal to the right of the selected cutset and search for a signal that satisfies the following criterion: *every one of its immediate fanin key signals is in the cutset.* Once a potential error source is located, back-substitution is performed in an attempt to fix it. Fig. 11.12 shows an example in which ε_2 is replaced by ε_1.

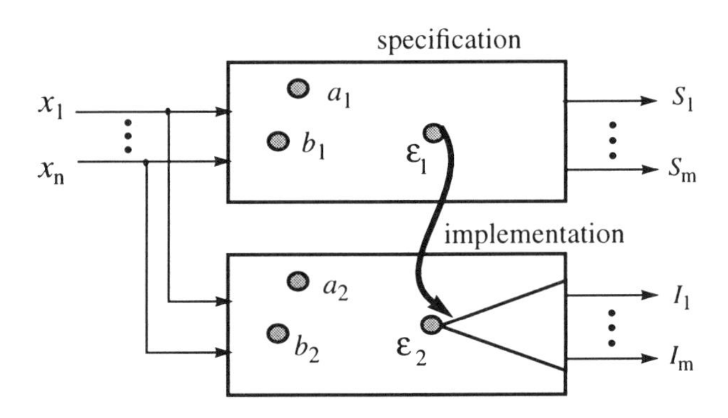

Fig. 11.12 Fix an error signal by back-substitution.

11.4 Experimental Results

A prototype tool, named AutoFix, has been implemented in C language in the SIS environment [121]. AutoFix uses three algorithms: (1) the symbolic approach (Section 11.2), (2) the hybrid approach for large circuits

(Section 11.3), and (3) the structural approach via back-substitution (Section 11.3.1).

The pure symbolic approach has been tested on a suite of 11 industrial engineering change examples from Fujitsu. In this experiment, we do not utilize any structural correspondence or naming information. The initial specification of each example is described in Verilog at the gate level, and the implementation is an optimized version of the old specification. The revised specification (due to engineering change) is obtained by modifying the functions of a number of internal signals in the initial Verilog specification. The results of AutoFix are shown in Table 11.1. The second column

Table 11.1 Engineering change results on a suite of industrial designs. (structural correspondence or naming information is **not** required)

designs	# gates (C_1/C_2)	rectified circuit (added, reused, total)	# partial fix	recycling-rate		rectification time (sec)	verification time (sec)
				[87] + [62]	new		
F1	217 / 217	(**0**, 217, 217)	0	100%	100%	-	2.8
F2	37 / 35	(**8**, 26, 34)	1	78.4%	77.8%	0.7	0.5
F3	133 / 132	(**4**, 130, 134)	1	94.7%	97.0%	3.0	1.8
F4	96 / 96	(**3**, 93, 96)	3	82.3%	96.9%	2.3	1.0
F5	818 / 534	(**8**, 530, 538)	4	63.4%	98.5%	61.8	39.6
F6	804 / 476	(**8**, 472, 480)	4	55.7%	98.3%	78.6	24.5
F7	465 / 301	(**8**, 299, 307)	2	61.3%	97.4%	20.4	13.1
F8	846 / 606	(**8**, 602, 610)	4	69.7%	98.7%	60.0	53.6
F9	645 / 465	(**8**, 462, 470)	3	70.2%	98.3%	42.2	27.0
F10	388 / 388	(**39**, 365, 404)	3	85.4%	90.4%	24.1	5.6
F11	363 / 344	(**19**, 343, 362)	1	87.5%	94.8%	57.8	38.2
Avg	-	-	-	77.1%	95.3%	-	-

shows the number of logic gates in the revised specification (C_1) and the initial implementation (C_2). The third column shows the statistics of the corrected implementation in terms of the number of newly added gates, the number of reused gates, and total number of gates. In this experiment, each partial correction generates a new sub-network attached to the initial implementation. This sub-network is then mapped into the library cells defined in *mcnc.genlib* to calculate the total number of newly added logic gates. For instance, the number of logic gates in the revised specification and the initial implementation for the design F6 are 804 and 476, respectively. AutoFix generates a final new implementation with 480 gates. Among these logic gates, only 8 have been newly added. The other 472 are the existing gates of the original implementation.

The fourth column shows the number of partial corrections required for each example. The revised specification and the initial implementation for the first example are proven equivalent and, thus, the implementation requires no correction. The number of partial corrections for other examples ranges from 1 to 4. It can be proven using *ErrorTracer* of Chapter 9, that for 5 out of 11 examples (F4, F5, F6, F8, F9, F10) at least *three* signals need to be re-synthesized in order to completely rectify the circuits (i.e., there exists no double-fix pair). This indicates that the strong partial-fix signals identified in AutoFix are good candidates for partially correcting the circuits. The fifth column shows the recycling rate, which is defined as the ratio of the number of reused gates to that of the total number of gates in the corrected implementation. Compared with the combined results of previously available programs, [62] and [87], the recycling rate is improved from 77% to 95% on an average. The sixth column shows the logic rectification time in seconds on a Sun Sparc-20 workstation. The functional equivalence of the corrected implementation and the revised specification has further been verified by AQUILA (Chapter 5). The verification time is reported in the last column.

Table 11.2 shows the results of the hybrid approach to fix the optimized ISCAS-85 benchmark circuits injected with one gate type error. The naming information is used to guide the correction process in this experiment. The initial implementations are obtained by optimizing the original benchmark circuits using *script.rugged* in SIS. We use a program [73] to randomly select one internal node and scramble its function. Column 2 shows the number of nodes in the erroneous implementation after being decomposed into AND and OR gates using SIS command "tech_decomp -a 5 -o 5". For 6 out of the 10 circuits, the implementation can be fixed by

Table 11.2 Results of correcting the optimized ISCAS-85 benchmark circuits with a single error. (Note that: the naming information is used to guide the correction process in this experiment)

circuits	# gates (C_1/C_2)	# single gate fix	# new gates by hybrid approach	rectification time (sec)	verification time (sec)
C432	123 / 49	YES	1	52	2
C499	370 / 378	YES	1	9	4
C880	302 / 266	NO	7*	5	3
C1355	474 / 378	YES	1	13	4
C1908	465 / 332	YES	1	41	19
C2670	694 / 584	YES	1	99	7
C3540	1020 / 1167	NO	9*	1844	475
C5315	1462 / 1118	YES	1	58	65
C6288	2353 / 2307	NO	5*	132	40
C7552	2118 / 2039	NO	8*	419	42

* indicates the number of the new gates by the pure structural approach using back-substitution.

modifying a single gate using the hybrid approach that performs symbolic correction in a divide-and-conquer way. The reason for AutoFix not being able to fix the other 4 circuits by only changing one gate is the heuristic of using the pseudo_PI set as the inputs and treating a selected cutset of structurally corresponding signal pairs as the outputs when performing the symbolic diagnosis. For these 4 circuits, the numbers of the new gates reported by the back-substitution based structural approach are listed.

Tables 11.3 and 11.4 show the results of the pure structural approach with back-substitution guided by the naming information. Table 11.3 shows the results of correcting the SIS-minimized circuits injected with 3 random gate-type errors. Table 11.4 shows the results of correcting the minimized C6288 with 1 to 10 injected errors.

Table 11.3 Results of correcting minimized circuits injected with 3 errors.

Circuit	# key signals	# new nodes	# existing nodes	Recycling rate	CPU-time (seconds)
C432	38	16	187	92%	8.0
C499	90	4	437	99%	6.7
C880	49	40	327	89%	12.1
C1355	90	6	447	99%	10.6
C1908	98	79	369	82%	31.4
C2670	208	27	918	97%	88.9
C3540	145	147	1167	89%	2025.0
C5315	292	105	1373	93%	27.0
C6288	704	14	2362	99%	35.8
C7552	398	190	2261	92%	42.4

Table 11.4 Results of correcting C6288 injected with multiple errors.

no. of errors	# key signals	# new nodes	# existing nodes	Recycling-rate	CPU-time (seconds)
1	704	3	2369	99.9%	89.8
2	704	7	2368	99.9%	89.8
3	704	12	2355	99%	94.9
4	704	8	2356	99.9%	102.1
5	704	36	2337	98%	114.3
6	704	16	2348	99%	93.5
7	704	35	2333	99%	96.8
8	704	31	2338	99%	97.8
9	704	36	2338	98%	105.6
10	704	127	2299	94%	140.9

11.5 Summary

Three logic rectification algorithms for the purposes of engineering change and error correction are discussed in this chapter. An interpretation of the necessary and sufficient condition for the single signal correctable condition is illustrated. Based on this interpretation, we introduce a generalized concept called partial correction. With this concept, we consider the logic rectification process consisting of a sequence of partial corrections, each reducing the set of the input vectors differentiating the specification and implementation. To improve the recycling rate and reduce the computational complexity, this algorithm employs efficient equivalence checking techniques to narrow down the error region. Also, the necessary and sufficient condition for changing only one single gate to realize a corrected function is derived. Experimental results for a suite of industrial examples demonstrate that on an average 95% of logic recycling can be achieved. Two ideas are presented to handle larger designs: (1) The incorporation of the structural correspondence between the specification and implementation into an enhanced symbolic approach, and (2) the incorporation of a simple heuristic for selecting a potential error source for performing back-substitution. With these enhancements, AutoFix can correct the entire set of optimized ISCAS-85 benchmark circuits injected with multiple random errors.

Bibliography

[1] M. S. Abadir, J. Ferguson, and T. E. Kirkland, "Logic Design Verification via Test Generation," *IEEE Trans. on Computers*, vol. 37, pp. 138-148 (Jan. 1988).

[2] M. Abramovici, M. A. Breuer, and A. D. Friedman, *Digital Systems Testing and Testable Design*, IEEE Press (1990).

[3] M. Abramovici, P. R. Menon, and D. T. Miller, "Critical Path Tracing: An Alternative to Fault Simulation," *IEEE Design & Test of Computers*, vol. 1, no. 1, pp. 83-93 (Feb. 1984).

[4] P. Agrawal, V. D. Agrawal, and K. T. Cheng, "Fault Simulation in a Pipelined Multiprocessor System," *Proc. Int't Test Conf.*, pp. 727-734 (Aug. 1989).

[5] V. D. Agrawal and D. Lee, "Characteristic Polynomial Method for Verification and Test of Combinational Circuits," *Proc. of 9th Int'l Conf. on VLSI Design*, pp. 341-342 (1996).

[6] V. D. Agrawal, D. Lee, and H. Wozniakowski, "Numerical Computation of Characteristic Polynomials of Boolean Functions and Its Applications," *Numerical Algorithms*, vol. 16 (1998).

[7] V. D. Agrawal and S. T. Chakradhar, "Combinational ATPG Theorems for Identifying Untestable Faults in Sequential Circuits," *IEEE Trans. on Computer-Aided Design*, vol. 14, pp. 1155-1160 (Sept. 1995).

[8] S. Akers, "Binary Decision Diagram," *IEEE Trans. on Computers*, vol. C-27, no. 6, pp. 509-516 (June 1978).

[9] H. Al-Asaad and J. P. Hayes, "Design Verification via Simulation and Automatic Test Pattern Generation," *Proc. of Int'l Conf. on Computer-Aided Design*, pp. 174-180 (Nov. 1995).

[10] P. Ashar, A. Gupta, and S. Malik, "Using Complete-1-Distinguishability for FSM Equivalence Checking", *Proc. of Int'l Conf. on Computer-Aided Design*, pp. 346-353 (Nov. 1996).

[11] R. A. Bergamaschi, D. Brand, L. Stok, M. Berkelaar, and S. Prakash, "Efficient Use of Large Don't Cares in High-Level and Logic Synthesis," *Proc. of Int'l Conf. on Computer-Aided Design*, pp. 272-278 (Nov. 1995).

[12] C. L. Berman and L. H. Trevillyan, "Functional Comparison of Logic Designs for VLSI Circuits," *Proc. of Int'l Conf. on Computer-Aided Design*, pp. 456-459 (Nov. 1989).

[13] D. Brand, "Verification of Large Synthesized Designs," *Proc. Int'l Conf. on Computer-Aided Design*, pp. 534-537 (Nov. 1993).

[14] D. Brand, A. Drumm, S. Kundu, and P. Narain, "Incremental Synthesis," *Proc. of Int'l Conf. on Computer-Aided Design*, pp. 14-18 (Nov. 1994).

[15] D. Brand, R. A. Bergamaschi, and L. Stok, "Be Careful With Don't Care," *Proc. of Int'l Conf. on Computer-Aided Design*, pp. 83-86 (Nov. 1995).

[16] R. K. Brayton et al. "VIS: Verification Interacting with Synthesis," *Proc. of Conf. on Computer-Aided Verification*, *Lecture Notes in Computer Science 1102*, pp. 408-412, Springer-Verlag (1996).

[17] F. M. Brown, *Boolean Reasoning*, Kluwer Academic Publisher (1990).

[18] R. E. Bryant, "Graph-based Algorithms for Boolean Function Manipulation," *IEEE Trans. on Computers*, vol. 35, No. 8, pp. 677-691 (Aug. 1986).

[19] R. E. Bryant and Y.-A. Chen, "Verification of arithmetic circuits with binary moment diagrams," *Proc. of Design Automation Conf.*, pp. 535-541 (June 1995).

[20] R. E. Bryant, D. Beatty, and C. Seger, "Formal Hardware Verification by Symbolic Ternary Trajectory Evaluation," *Proc. of Design Automation Conf.*, pp. 397-402 (June 1991).

[21] J. Burch, E. Clark, D. Long, K. McMillan, and D. Dill, "Symbolic Model Checking for Sequential Circuit Verification," *IEEE Trans. on Computer-Aided Design*, vol. 13, no. 4, pp. 401-424 (April 1994).

[22] M. L. Bushnell and J. Giraldi, "A Functional Decomposition Method for Redundancy Identification and Test Generation," *Journal of*

Electronic Testing: Theory and Applications, vol. 10, pp. 175-195 (June 1997).

[23] M. Butts, J. Batcheller, and J. Varghese, "An Efficient Logic Emulation System," *Proc. Int'l Conf. on Computer Design (ICCD-92)*, pp. 138-141 (Oct. 1992).

[24] S. T. Chakradhar, V. D. Agrawal, and M. L. Bushnell, "Neural Net and Boolean Satisfiability Models of Logic Circuits," *IEEE Design & Test of Computers*, vol. 7, pp. 54-57 (Oct. 1990).

[25] S. T. Chakradhar, V. D. Agrawal, and M. L. Bushnell, *Neural Models and Algorithms for Digital Testing*, Kluwer Academic Publishers (1991).

[26] S. T. Chakradhar, M. L. Bushnell, and V. D. Agrawal, "Toward Massively Parallel Automatic Test Generation," *IEEE Trans. on Computer-Aided Design*, vol. 9, pp. 981-994 (Sept. 1990).

[27] S. Chang, D. I. Cheng, and M. Marek-Sadowska, "Minimizing ROBDD Size of Incompletely Specified Multiple Output Functions," *Proc. of European Design and Test Conf.*, pp. 620-624 (1994).

[28] S. Chang and Marek-Sadowska, "Perturb and Simplify: Multi-level Boolean Network Optimizer," *Proc. of Int'l Conf. on Computed-Aided Design*, pp. 2-5 (Nov. 1994).

[29] X. Chen and M. L. Bushnell, *Efficient Branch and Bound Search with Application to Computer-Aided Design*, Kluwer Academic Publishers (1996).

[30] K.-T. Cheng and V. D. Agrawal, *Unified Methods for VLSI Simulation and Test Generation*, Kluwer Academic Publishers (1989).

[31] K.-T. Cheng and V. D. Agrawal, "Initializability Considerations in Sequential Machine Synthesis," *IEEE Trans. on Computers*, vol. 41, pp. 374-379 (March 1992).

[32] K.-T. Cheng, "Redundancy Removal for Sequential Circuits Without Reset States," *IEEE Trans. on Computer-Aided Design*, pp. 652-667 (Jan. 1993).

[33] K.-T. Cheng and L. A. Entrena, "Multi-level logic optimization by redundancy addition and removal," *Proc. of European Conf. on Design Automation*, pp. 373-377 (June 1993).

[34] K.-T. Cheng and H.-K T. Ma, "On the Over-specification Problem in Sequential ATPG Algorithms," *IEEE Trans. on Computer-Aided Design*, pp. 1599-1604 (Oct. 1993).

[35] W.-T. Cheng, "The BACK Algorithm for Sequential Test Generation," *Proc. Int'l Conf. on Computer Design*, pp. 66-69 (Oct. 1988).

[36] W.-T. Cheng and M.-L. Yu, "Differential Fault Simulation - A Fast Method Using Minimal Memory," *Proc. of Design Automation Conf.*, pp. 424-428 (June 1989).

[37] H. Cho, G. D. Hachtel, S.-W. Jeong, B. Plessier, E. Schwarz, and F. Somenzi, "ATPG Aspects of FSM verification," *Proc. of Int'l Conf. on Compute-Aided Design*, pp. 134-137 (Nov. 1990).

[38] H. Cho and F. Somenzi, "Sequential Logic Optimization Based on State Space Decomposition," *Proc. of European Conf. on Design Automation*, pp. 200-204 (Feb. 1993).

[39] H. Cho, S.-W. Jeong, F. Somenzi and C. Pixley, "Synchronizing Sequences and Symbolic Traversal Techniques in Test Generation," *Journal of Electronic Testing: Theory and Applications*, vol. 4, no. 12, pp. 19-31 (1993).

[40] P.-Y. Chung, Y. M., Wang, and I. N., Hajj, "Logic design error diagnosis and correction", *IEEE Transactions on VLSI Systems*, vol. 2, no. 3, pp. 320-332 (Sept. 1994).

[41] E. Clarke, E. A. Emerson, and A. Sistla, "Automatic Verification of Finite State Concurrent Systems Using Temporal Logic Specifications," *ACM Trans. on Programming Language and Systems*, vol. 1, no. 2, pp. 244-263 (April 1986).

[42] O. Coudert, C. Berthet and J. C. Madre, "Verification of Synchronous Sequential Machines Based on Symbolic Execution," *Automatic Verification Methods for Finite State System, LNCS no. 407*, Springer Verlag (1990).

[43] O. Coudert and J. C. Madre, "A Unified Framework for the Formal Verification of Sequential Circuits," *Proc. of Int'l Conf. on Computed-Aided Design*, pp. 126-129 (1990).

[44] S. Devadas, H.-K. T. Ma, and A. Sangiovanni-Vincentelli, "Logic Verification, Testing and Their Relationship to Logic Synthesis,"

Testing & Diagnosis of VLSI & ULSI, Kluwer Academic Publishers, pp. 181-246 (1988).

[45] P. A. Duba, R. K. Roy, J. A. Abraham, and W. A. Rogers, "Fault Simulation in a Distributed Environment," *Proc. of Design Automation Conf.*, pp. 686-691 (June 1988).

[46] L. Entrena and K.-T. Cheng, "Sequential Logic Optimization by Redundancy Addition and Removal," *Proc. Int'l Conf. on Computer-Aided Design*, pp. 310-315 (Nov. 1993).

[47] M. Fujita, Y. Tamiya, Y. Kukimoto, and K.-C. Chen, "Application of Boolean Unification to Combinational Logic Synthesis," *Proc. of Int'l Conf. on Computer-Aided Design*, pp. 510-513, Nov. 1991.

[48] M. Fujita, "Methods for Automatic Design Error Correction in Sequential Circuits," *Proc. of European Conf. on Design Automation*, pp. 76-80 (1993).

[49] H. Fujiwara, T. Shimono, "On the Acceleration of Test Generation Algorithms", *Proc. of 13-th Fault Tolerant Computing*, pp. 98-105, 1983.

[50] A. Ghosh, S. Devadas and A. R. Newton, "Test Generation and Verification for Highly Sequential Circuits," *IEEE Trans. on Computer-Aided Design,* vol. 10, pp. 652-667 (May 1991).

[51] P. Girard, C. Landrault, and S. Pravossoudovitch, "Delay Fault Diagnosis by Critical-Path Tracing," *IEEE Design and Test of Computers*, pp. 27-32 (Dec. 1992).

[52] U. Glaeser and H. T. Vierhaus "Mixed Level Test Generation for Synchronous Sequential Circuits using the FOGBUSTER-Algorithm," *IEEE Trans on Computer-Aided Design*, vol. 15, No. 4, pp. 410 - 423 (April 1996).

[53] N. Gouders, R. Kaibel, "PARIS: A Parallel Pattern Fault Simulator for Synchronous Sequential Circuits," *Proc. Int'l Conf. on Computer-Aided-Design*, pp. 542-545 (Nov. 1991).

[54] P. Goel, "An Implicit Enumeration Algorithm to Generate Tests for Combinational Logic Circuits," *IEEE Trans. on Computers*, vol. C-30, no. 3, pp., 215-222 (March 1981).

[55] A. Gupta, "Formal Hardware Verification Methods: A Survey," *Formal Methods in System Design*, vol. 1, No. 2, pp. 151-238 (Oct. 1992).

[56] J. P. Hayes, "*Computer architecture and organization*," 2nd edition, McGraw-Hill (1988).

[57] G. J. Holzmann, *Design and Validation of Computer Protocols*, Prentice Hall (1991).

[58] Y. Hong, P. A. Beerel, J. R. Burch, and K. L. McMillan, "Safe BDD Minimization Using Don't Cares," *Proc. of Design Automation Conf.*, pp. 208-213 (June, 1997).

[59] Y. V. Hoskote, "*Formal Techniques for Verification of Synchronous Sequential Circuits*," Ph.D. Thesis, Univ. of Texas at Austin (Dec. 1995).

[60] S.-Y. Huang, K.-T. Cheng, K.-C. Chen, and U. Glaeser, "An ATPG-Based Framework for Verifying Sequential Equivalence," *Proc. of Int'l Test Conf.*, pp. 865-874 (Oct. 1996).

[61] S.-Y. Huang, K.-T. Cheng, and K.-C. Chen, "On Verifying the Correctness of Retimed Circuits," *Proc. of Great-Lake Symposium on VLSI*, pp. 277-280 (March 1996).

[62] S.-Y. Huang, K.-C. Chen, and K.-T. Cheng, "Error Correction Based on Verification Techniques," *Proc. of Design Automation Conf.*, pp. 258-261 (June 1996).

[63] S.-Y. Huang, K.-T. Cheng, and K.-C. Chen, "AQUILA: An Equivalence Verifier for Large Sequential Circuits," *Proc. of Asia and South Pacific Design Automation Conf.*, pp. 455-460 (Jan. 1997).

[64] S.-Y. Huang, K.-C. Chen, and K.-T. Cheng, "Incremental Logic Rectification," *Proc. of VLSI Test Symposium*, pp. 134-139 (April 1997).

[65] S.-Y. Huang, K.-T. Cheng, K.-C. Chen, and D. I. Cheng, "Error-Tracer: A Fault Simulation Based Approach to Design Error Diagnosis," *Proc. of Int'l Test Conf.*, pp. 974-981 (Nov. 1997).

[66] S.-Y. Huang, K.-T. Cheng, K.-C. Chen, and J.-Y. Lu, "Fault Simulation Based Design Error Diagnosis for Sequential Circuits," *Proc. of Design Automation Conf.* (June 1998).

[67] W. Hunt, "Microprocessor Design Verification," *Journal of Automated Reasoning*, vol. 5(4), pp. 429-460 (Dec. 1989).

[68] V. S. Iyengar and D.-T. Tang, "On Simulation Faults in Parallel," *Digest of Papers of 18th Int'l Symposium on Fault-Tolerant Computing*, pp. 110-115 (June 1988).

[69] M. A. Iyer, "*On Redundancy and Untestability in Sequential Circuits*," Ph.D. Thesis, Department of ECE, Illinois Institute of Technology, Chicago, Illinois (July 1995).

[70] J. Jain, J. Bitner, D. S. Fussel, and J. A. Abraham, "Probabilistic Verification of Boolean Formulas," *Formal Methods in System Designs*, vol. 1, pp. 63-117 (1992).

[71] J. Jain, R. Mukherjee, and M. Fujita, "Advanced Verification Techniques Based on Learning," *Proc. of Design Automation Conf.*, pp. 420-426 (June 1995).

[72] F. Krohm, A. Kuehlmann, and A. Mets, "The Use of Random Simulation in Formal Verification," *Proc. of Int'l Conf. on Computer Design* (Oct. 1996).

[73] A. Kuehlmann, D. I. Cheng, A. Srinivasan, and D. P. LaPotin, "Error Diagnosis for Transistor-level verification," *Proc. of Design Automation Conf.*, pp. 218-223 (June 1994).

[74] A. Kuehlmann, A. Srinivasan, and D. P. LaPotin, "Verity - A Formal Verification Program for Custom CMOS Circuits," *IBM Journal of Research and Development*, vol. 39, pp.149-165 (1995).

[75] A. Kuehlmann and F. Krohm, "Equivalence Checking Using Cuts and Heaps," *Proc. of Design Automation Conf.*, pp. 263-268 (June, 1997).

[76] Y. Kukimoto and M. Fujita, "Rectification Method for Look-up Type FPGA's", *Proc. of Int'l Conf. on Computer-Aided Design*, pp. 54-61 (Nov. 1992).

[77] W. Kunz and D. K. Pradhan, "Recursive Learning: An Attractive Alternative to the Decision Tree for Test Generation in Digital Circuits," *Proc. of Int'l Test Conf.*, pp. 816-825, 1992.

[78] W. Kunz, "HANNIBAL: An Efficient Tool for Logic Verification Based on Recursive Learning," *Proc. of Int'l Conf. on Computer-Aided Design*, pp. 538-543 (Nov. 1993).

[79] W. Kunz and D. Stoffel, *Reasoning in Boolean Networks: Logic Synthesis and Verification Using Testing Techniques*, Kluwer Academic Publishers (1997).

[80] S.-Y. Kuo, "Locating Logic Design Errors via Test Generation," *Proc. of European Design Automation Conf.*, pp. 466-467 (1992).

[81] R. P. Kurshan, *Computer-Aided Verification of Coordinating Process*, Princeton University Press, Princeton, NJ (1994).

[82] T. Larrabee, "Test Generation Using Boolean Satisfiability," *IEEE Trans. on Computer-Aided Design*, vol. 11, pp. 4-15 (Jan. 1992).

[83] D. H. Lee and S. M. Reddy, "On Efficient Simulation For Synchronous Sequential Circuits", *Proc. 29th Design Automation Conf.*, pp. 327-331 (June 1992).

[84] H. K. Lee and D. S. Ha, "HOPE: An Efficient Parallel Fault Simulator for Synchronous Sequential Circuits," *Proc. 29th Design Automation Conf.*, pp. 336-340 (June 1992).

[85] H. K. Lee and D. S. Ha, "New Methods of Improving Parallel Fault Simulation in Synchronous Sequential Circuits," *Proc. of Int'l Conf. on Computer-Aided- Design*, pp. 10-17 (Nov. 1993).

[86] H.-T. Liaw, J.-H. Tsaih, and C.-S. Lin, "Efficient Automatic Diagnosis of Digital Circuits", *Proc. of Int'l Conf. on Computer-Aided Design*, pp. 464-467 (Nov. 1990).

[87] C.-C. Lin, K.-C. Chen, S.-C. Chang, M. Marek-Sadowska, and K.-T. Cheng, "Logic Synthesis for Engineering Change," *Proc. of Design Automation Conf.*, pp. 647-652 (June 1995).

[88] C.-C. Lin, K.-C. Chen, D. I. Cheng, and M. Marek-Sadowska, "Logic Rectification and Synthesis for Engineering Change," *Proc. of Asia South Pacific Design Automation Conf.*, pp. 301-309 (Jan. 1995).

[89] C. Y. Lo, H. N. Nham, and A. K. Bose, "Algorithms for an Advanced Fault Simulation System in Motis," *IEEE Trans. on Computer-Aided Design*, vol. CAD-6, pp. 232-240 (March 1987).

[90] C. E. Leiserson and J. B. Saxe, "Retiming Synchronous Circuitry," *Algorithmica*, vol. 6, pp. 5-35 (1991).

[91] K. L. McMillan, *Symbolic Model Checking*, Kluwer Academic Publisher, Boston, MA (1994).

[92] J. C. Madre, O. Coudert, and J. P. Billon, "Automating the Diagnosis and the Rectification of the Design Errors with PRIAM," *Proc. of Int'l Conf. on Computer-Aided Design*, pp. 30-33 (Nov. 1989).

[93] S. Malik, E. M. Santovich, R. K. Brayton, and A. Sangiovanni-Vincentelli, "Retiming and Resynthesis: Optimizing Sequential Networks with Combinational techniques," *IEEE Trans. on Computer-Aided Design*, vol. 10, pp. 74-84 (Jan. 1991).

[94] T. E. Marchok, A. El-Maleh, J. Rajski, and W. Maly, "Test Set Preservation under Retiming Transformation," *Proc. of Design Automation Conf.*, pp. 414-419 (June 1995).

[95] R. Marlett, "EBT: A Comprehensive Test Generation Technique for Highly Sequential Circuits," *Proc. of 15th Design Automation Conf.*, pp. 332-339 (June 1978).

[96] U. Martin and T. Nipkow, "Boolean Unification - The Story So Far," *Journal of Symbolic Computation*, vol. 7, pp. 275-293, 1989.

[97] Y. Matsunaga, "An Efficient Equivalence Checker for Combinational Circuits," *Proc. of Design Automation Conf.*, pp. 629-634 (June 1996).

[98] E. J. McCluskey, *Introduction to the Theory of Switching Circuits*, New York, McGraw-Hill (1965).

[99] R. Murgai, N. Shenoy R. K. Brayton, and A. Sangiovanni-Vincentelli. "Improved Logic Synthesis Algorithms for Table Look Up Architecture." *Proc. of Int'l Conf. on Computer-Aided Design*, pp. 564-567 (Nov. 1991).

[100] S. Muroga, Y. Kambayashi, H. C. Lai and J. N. Culliney, "The Transduction Method - Design of Logic Networks Based on Permissible Functions," *IEEE Trans. on Computers*, vol. 38, pp. 1404-1424 (1989).

[101] V. Narayanan and V. Pitchumani, "A Massively Parallel Algorithm for Fault Simulation on the Connection Machine," *Proc. of 26th Design Automation Conf.*, pp. 734-737 (June 1989).

[102] T. M. Niermann, W. T. Cheng, and J. H. Patel, "PROOFS: A Fast and Memory Efficient Sequential Circuit Fault Simulator," *IEEE Trans. on Computer-Aided Design*, vol. 11, pp. 198-207 (Feb. 1992).

[103] D. L. Ostapko, Z. Barzilai, and G. M. Silberman, "Fast Fault Simulation in a Parallel Processing Environment," *Proc. of Int'l Test Conf.*, pp. 686-691 (Sept. 1987).

[104] C. Pixley, "A Theory and Implementation of Sequential Hardware Equivalence", *IEEE Trans. on Computer-Aided Design*, pp. 1469-1494 (Dec. 1992).

[105] C. Pixley, V. Singhal, A. Aziz, and R. K. Brayton, "Multi-level Synthesis for Safe Replaceability," *Proc. of Int'l Conf. on Computer-Aided Design*, pp. 442-449 (Nov. 1994).

[106] I. Pomeranz and S. M. Reddy, "A method for diagnosing implementation errors in synchronous sequential circuits and its implications on synthesis," *Proc. of European Design Automation Conf.*, pp. 252-258 (Sept. 1993).

[107] I. Pomeranz and S. M. Reddy, "On Achieving Complete Testability of Synchronous Sequential Circuits with Synchronizing Sequences," *Proc. of Int'l Test Conf.*, pp. 1007-1016 (Oct. 1994).

[108] I. Pomeranz and S. M. Reddy, "On Error Correcting in Macro-based Circuits," *Proc. of Int'l Conf. on Computer-Aided Design*, pp. 568-575 (Nov. 1994).

[109] I. Pomeranz and S. M. Reddy, "On Correction of Multiple Design Errors," *IEEE Trans. on Computer-Aided Design*, vol. 14, no. 2, pp. 255-264 (Feb. 1995).

[110] D. K. Pradhan, D. Paul, and M. Chatterjee, "VERILAT: Verification Using Logic Augmentation and Transformations," *Proc. of Int'l Conf. on Computer-Aided Design*, pp. 88-95 (Nov. 1996).

[111] Quick-Turn Design System Inc., "MARSIII Emulation System User's Guide" (Jan. 1994).

[112] S. M. Reddy, W. Kunz, and D. K. Pradhan, "Novel Verification Framework Combining Structural and OBDD Methods in a Synthesis Environment," *Proc. of Design Automation Conf.*, pp. 414-419 (June 1995).

[113] J. P. Roth, *Computer Logic, Testing, and Verification*, Computer Science Press (1980).

[114] S. Rudeanu, *Boolean Functions and Equations*, North-Holland Publishing Co. Amsterdam (1974).

[115] R. Rudell, "Dynamic Variable Ordering for Ordered Binary Decision Diagram," *Proc. of Int'l Conf. on Computer-Aided Design*, pp. 42-47 (Nov. 1993).

[116] N. Shenoy and R. K. Brayton, "Retiming of Circuits with Single phase transparent latches," *Proc. of Int'l Conf. on Computer Design*, pp. 86-89 (Nov. 1991).

[117] N. Shenoy and R. Rudell, "Efficient Implementation of Retiming," *Proc. Int'l Conf. on Computer-Aided Design*, pp. 226-233 (Nov. 1994).

[118] M. Singh and S. M. Nowick, "Synthesis for Logical Initializability of Synchronous Finite State Machine," *Proc. of 10th Int'l Conf. on VLSI Design*, pp. 76-80 (1997).

[119] V. Singhal, C. Pixley, R. Rudell, R. K. Brayton, "The Validity of Retiming Sequential Circuits," *Proc. of 32th Design Automation Conf.*, pp. 316-321 (June 1995).

[120] T. Shiple, R. Hojati, A. Sangiovanni-Vincentelli, and R. K. Brayton, "Heuristic Minimization of BDDs Using Don't Cares", *Proc. of Design Automation Conf.*, pp. 225-231 (June, 1994).

[121] "SIS: A System for Sequential Circuit Synthesis," Report M92/41, University of California, Berkeley (May 1992).

[122] N. C. E. Srinivas and V. D. Agrawal, "Formal Verification of Digital Circuits Using Hybrid Simulation," *IEEE Circuits and Devices*, vol. 4, pp. 19-27 (Jan. 1988).

[123] K. A. Tamura, "Locating Functional Errors in Logic Circuits," *Proc. of Design Automation Conf.*, pp. 185-191 (June 1989).

[124] H. J. Touati, H. Sarvoj, B. Lin, R. K. Brayton and A. Sangiovanni-Vincentelli, "Implicit State Enumeration of Finite State Machines Using BDD's," *Proc. Int'l Conf. on Computer-Aided Design*, pp. 130-133 (Nov. 1990).

[125] E. G. Ulrich, V. D. Agrawal, and J. H. Arabian, *Concurrent and Comparative Discrete Event Simulation*, Kluwer Academic Publishers (1994).

[126] A. G. Veneris and I. N. Hajj, "A Fast Algorithm for Locating and Correcting Simple Design Errors in VLSI Digital Circuits," *Proc. of Great Lake Symposium on VLSI Design*, pp. 45-50 (March 1997).

[127] A. M. Wahba and D. Borrione, "A method for automatic design error location and correction in combinational logic circuits", *Journal of Electronic Testing: Theory and Applications*, vol.8, no.2, pp. 113-27 (April 1996).

[128] A. M. Wahba and D. Borrione, "Design Error Diagnosis in Sequential Circuits," *Proc. of Correct Hardware Designs and Verification Methods*, CHARME'95, Lecture Notes in Computer Science, no. 987, pp. 171-188, Springer Verlag (Oct. 1995).

[129] S. Walters, "Computer-Aided Prototyping for ASIC-Based System," *IEEE Design & Test of Computer*, pp. 4-10 (June 1991).

[130] Y. Watanabe and R. K. Brayton, "Incremental synthesis for Engineering Change," *Proc. of Int'l Conf. on Computer-Aided Design*, pp. 40-43 (Nov. 1991).

Index

D

E

T